Handbook of CCD Astronomy, Second edition.

Charge-coupled devices (CCDs) are the state-of-the-art detectors in many fields of observational science. Updated to include all of the latest developments in CCDs, this second edition of the *Handbook of CCD Astronomy* is a concise and accessible reference on all practical aspects of using CCDs. Starting with their electronic workings, it discusses their basic characteristics and then gives methods and examples of how to determine these values.

While the book focuses on the use of CCDs in professional observational astronomy, advanced amateur astronomers and researchers in physics, chemistry, medical imaging, and remote sensing will also find it very valuable. Tables of useful and hard-to-find data, key practical equations, and new exercises round off the book and ensure that it provides an ideal introduction to the practical use of CCDs for graduate students, and a handy reference for more experienced users.

STEVE HOWELL is an astronomer at the National Optical Astronomical Observatory and WIYN Observatory in Tucson, Arizona. He began working at Kitt Peak National Observatory in the early 1980s as a support scientist for the then brand new CCD systems being put into service at the observatory. Since then he has worked on a number of interplanetary missions as part of their CCD imager teams, and built instruments for the NASA Space Shuttle. He is currently involved in research on interacting binary stars and ultra-high precision photometry and optical images using new technology CCDs.

Cambridge Observing Handbooks for Research Astronomers

Today's professional astronomers must be able to adapt to use telescopes and interpret data at all wavelengths. This series is designed to provide them with a collection of concise, self-contained handbooks, which covers the basic principles peculiar to observing in a particular spectral region, or to using a special technique or type of instrument. The books can be used as an introduction to the subject and as a handy reference for use at the telescope, or in the office.

Series editors
Professor Richard Ellis, Institute of Astronomy, *University of Cambridge*
Professor John Huchra, Center for Astrophysics, *Smithsonian Astrophysical Observatory*
Professor Steve Kahn, Department of Physics, *Columbia University*, New York
Professor George Rieke, Steward Observatory, *University of Arizona*, Tucson
Dr Peter B. Stetson, Herzberg Institute of Astrophysics, *Dominion Astrophysical Observatory*, Victoria, British Columbia

Handbook of CCD Astronomy

Second edition

STEVE B. HOWELL
National Optical Astronomy Observatory
and WIYN Observatory

CAMBRIDGE
UNIVERSITY PRESS

CAMBRIDGE UNIVERSITY PRESS
Cambridge, New York, Melbourne, Madrid, Cape Town, Singapore, São Paulo
Delhi, Tokyo, Mexico City

Cambridge University Press
The Edinburgh Building, Cambridge CB2 8RU, UK

Published in the United States of America by Cambridge University Press, New York

www.cambridge.org
Information on this title: www.cambridge.org/9780521852159

First published 2000
Second edition 2006
3rd printing 2011

Printed in the United Kingdom at the University Press, Cambridge

A catalog record for this publication is available from the British Library

ISBN 978-0-521-85215-9 Hardback
ISBN 978-0-521-61762-8 Paperback

Contents

Preface to the first edition

We are all aware of the amazing astronomical images produced with telescopes these days, particularly those displayed as color representations and shown off on websites and in magazines. For those of us who are observers, we deal with our own amazing images produced during each observing run. Just as spectacular are photometric, astrometric, and spectroscopic results generally receiving less fanfare but often of more astrophysical interest. What all of these results have in common is the fact that behind every good optical image lies a good charge-coupled device.

Charge-coupled devices, or CCDs as we know them, are involved in many aspects of everyday life. Examples include video cameras for home use and those set up to automatically trap speeders on British highways, hospital X-ray imagers and high-speed oscilloscopes, and digital cameras used as quality control monitors. This book discusses these remarkable semiconductor devices and their many applications in modern day astronomy.

Written as an introduction to CCDs for observers using professional or off-the-shelf CCD cameras as well as a reference guide, this volume is aimed at students, novice users, and all the rest of us who wish to learn more of the details of how a CCD operates. Topics include the various types of CCD; the process of data taking and reduction; photometric, astrometric, and spectroscopic methods; and CCD applications outside of the optical bandpass. The level of presentation was aimed not only at college or professional level readers but also at a more general audience including the ever-growing number of highly trained and motivated amateurs and other professionals in technical areas in which CCDs play a role.

Chapters 2 and 3 contain all the fundamental information on CCD operation and characteristics while each remaining chapter can be mastered individually. In a book of this length, many aspects must be treated briefly. However, I have made an effort to provide self-contained detail of the important aspects of CCDs while including numerous references to the detailed professional literature for those desiring a deeper understanding. Additionally, throughout the book, examples related to common observational occurrences as well as footnotes discussing interesting but not mainstream topics are included. Appendices list other reference works of interest, CCD manufacturers, numerous website addresses of interest, and a brief introduction to image displays.

This book started with an idea for a new series of handbooks, including a volume on CCDs. I am happy to thank the editor, Adam Black, and Peter Stetson for allowing me to be involved in this series and write this book. The folks at Cambridge University Press, particularly Adam, have been very helpful, dealing with my many questions during the writing process. Michie Shaw and the staff at TechBooks have helped greatly in the final steps of production. I would like to thank the anonymous readers of an early draft of this book for their comments and for pointing out some important areas and results I had overlooked. Two readers, Peter Stetson and John Huchra, suggested coverage of material that has led to the inclusion of additional topics in the final version. A number of colleagues have provided information, graphs, references, and support during the writing of this book, all of which I appreciate. I thank my former and current students and postdocs for keeping me on my toes regarding many of the topics herein. Chris Sabbey kindly provided Figure 6.8, and the color figures in the book (Figs. 1.1 and 4.6) were taken by Simon Tulloch and provided by Derek Ives, both of the UK Astronomy Technology Centre.

I would like to acknowledge and thank my parents, Cecil and Barbara, for allowing me to be "scientific" during my childhood. My experimentation, often at the expense of household items, was not always successful but was never discouraged. These experiences deeply planted the seed of scientific fascination in my being. Appreciation is also passed along to my brother Terry, for the many hours we spent together exploring the world around us. Particularly noteworthy were the times we spent watching, analyzing, and laughing at "B" sci-fi movies.

During the writing of this volume on CCDs, many opportunities were missed related to spending time with my son Graham and my wife and friend Mary. Both were always supportive of the effort, encouraged its completion, and have accumulated many IOUs, which I will now have the pleasure of paying off. I appreciate their unfailing love.

I have had fun writing this book and learning even more about CCDs, almost as much fun as I have when I observe with them. I hope that you the reader will find this work of interest as well and enjoy paging through it often. Astronomy has always fascinated humans and if this treatise allows you to obtain a better knowledge of CCDs and with it even more fascination with the Universe around us, it will have been a success.

"Since The Beginning Of Time, The Universe Has Called To Awaken Each Of Us. To Understand The Universe Is To Understand Ourselves."

Preface to the second edition

Seven years ago, Cambridge University Press began a new series of books called Handbooks. I was fortunate enough to be asked to author the one on CCDs. Little did I realize how wonderful of an undertaking that writing this book would be. I have learned and relearned a number of details about CCDs and had cause to read many scientific and popular papers and articles I otherwise would have overlooked. The greatest benefit, however, has been the many gracious colleagues and students who have provided comments, revisions, suggestions, support, and simply said thanks. The first edition of the *Handbook of CCD Astronomy* was written for you and you have truly made it your own through this volume.

When I was first asked to write a second edition, I have to admit I was skeptical that enough had changed to warrant it. I am happy to say I was completely wrong. Upon going back and reading the original volume, I had no problem seeing its many pages of outdated material. There are, however, some fundamental discussions and properties of CCDs that are timeless, and remain in the present volume. New areas of CCD development abound and to highlight a few this second edition is a bit longer and has a few more illustrations. The areas of faster and higher performance electronics to control and read out a CCD, better analog-to-digital circuitry, and better manufactured CCDs are some of the additions discussed within. The largest advance since the first edition is the continued development of new types of CCD. In the first edition, it was remarked that since CCDs have quantum efficiencies of nearly 80% or more, increases in this area will have little impact in overall performance. Today's CCD makers are providing surprising additional advances such as output registers with gain stages to amplify weak signals *in situ*, or the ability to manipulate the shape and location of the collected change during an exposure. These "wafer level" design changes, output speed and reliability, and a new generation of instruments and telescopes has provided a renaissance in CCD astronomy.

I would like to thank the many students of astronomy for their kind acceptance of this volume and the equally kind words of support they have provided. Abhishek Rawat and Joe Harrington found a few typos and provided specific comments leading to some clarifications in presentation. Three anonymous reviewers examined the first edition and my suggestions for changes, additions, and deletions to it. They provided sound guidance for the present volume. At Cambridge University Press, Simon Mitton has been a constant supporter and his experience and kind words have helped this project reach completion. Jacqueline Garget has kindly answered many questions and orchestrated the process for the second edition with alacrity. Vince Higgs and Tom Dolan have dealt with the detailed day-to-day issues of illustrations and page limits. I also wish to thank Wendy Phillips and my copy editor Karen Sawyer for their hard work and many explanatory emails. John Feldmeier and Jill Gerke were a great help in proofing the book and Elain Owens, once again, helped me produce the index. I thank them all for their continued kindness.

While writing the predecessor to the current book, my son Graham was a teenager just beginning to drive cars and still a boy in many ways. Now, a few cars later, he is a grown man and has a life of his own but still teaches me many things about how to live a good life and be a whole person. I am writing this preface on the eve of his birthday. My wife Mary is a constant inspiration to me as she always seems to know how to "do the life thing" much to my chagrin and amazement. She still shares her "Mary money" with me. Thanking my family, both direct relatives and those who have adopted me, can not express my full appreciation to them. They have supported me during this undertaking as well as in everyday life.

I hope you, the reader, will view this book as a starting point for your exploration and relationship with CCDs and astronomical detectors of all sorts. I encourage you to try them out in person as often as you can and pursue their use particularly in ways no one else has thought of. This is how astronomy and science advances and you are the future, our future. Enjoy.

1

Introduction

Silicon. This semiconductor material certainly has large implications on our life. Its uses are many, including silicon oil lubricants, implants to change our bodies' outward appearance, electric circuitry of all kinds, nonstick frying pans, and, of course, charge-coupled devices.

Charge-coupled devices (CCDs) and their use in astronomy will be the topic of this book. We will only briefly discuss the use of CCDs in commercial digital cameras and video cameras but not their many other industrial and scientific applications. As we will see, there are four main methods of employing CCD imagers in astronomical work: imaging, astrometry, photometry, and spectroscopy. Each of these topics will be discussed in turn. Since the intrinsic physical properties of silicon, and thus CCDs, are most useful at optical wavelengths (about 3000 to 11 000 Å), the majority of our discussion will be concerned with visible light applications. Additional specialty or lesser-used techniques and CCD applications outside the optical bandwidth will be mentioned only briefly. The newest advances in CCD systems in the past five years lies in the areas of (1) manufacturing standards that provide higher tolerances in the CCD process leading directly to a reduction in their noise output, (2) increased quantum efficiency, especially in the far red spectral regions, (3) new generation control electronics with the ability for faster readout, low noise performance, and more complex control functions, and (4) new types of scientific grade CCDs with some special properties. These advances will be discussed throughout the text.

Applications of infrared (IR) arrays – semiconductor devices with some similar properties to CCDs – while important to modern astronomy, will receive small mention here. A complete treatment of IR arrays is given by Ian Glass in his companion book, *Handbook of Infrared Astronomy*, the first book in this series.

Appendix A provides the reader with a detailed list of other books and major works devoted to CCD information. This appendix does not contain the

1

large body of journal articles in existence; these will be selectively referenced throughout the text. Appendix B provides a list of present day CCD manufacturers that produce both amateur and professional grade CCDs and CCD cameras, observatory websites, and other websites that contain useful CCD information. Finally, Appendix C discusses some basic principles of image display devices. While not directly related to CCDs, computer displays are the medium by which we view and examine the information collected by CCDs. Proper interpretation of the CCD image is only possible if one understands how it is displayed.

1.1 Nomenclature

CCDs are often listed and named by an apparently strange convention. This small section aims to demystify these odd sounding names. CCDs come in various sizes and shapes and are manufactured by a number of companies (see Appendix B).

Figure 1.1 illustrates a number of modern CCDs. Present day CCDs generally come in sizes ranging from 512 by 512 "picture elements" or pixels to arrays as large as 8192 by 8192 pixels. Often the size designations of CCDs are written as 2048×2048 or 2048^2. CCDs are also available as rectangular devices of unequal length and width and with nonsquare pixels. For example, CCDs of size 2048×4096 pixels are produced for spectroscopic applications. We will see in Chapter 2 that each pixel acts as an electrically isolated portion of the silicon array and is capable of incoming photon collection, storage of the produced photoelectrons, and readout from the CCD array to an associated computer as a digital number.

The names or designations of CCDs are usually a combination of the company name with the CCD size. Tek2048, $4K \times 2K$ E2V, and SITe4096 are examples. Instrumentation at observatories almost exclusively includes a CCD as the detector and is specialized to perform a task such as imaging or spectroscopy. Observatories designate these instruments with a name that may or may not include information about the associated CCD detector. The Royal Greenwich Observatory (RGO) on La Palma has the FOS#1A (a 512×1024 Loral CCD used in their Faint Object Spectrograph), and the Tek 2K CCD of the National Optical Astronomy Observatories (NOAO) is a 2048×2048 pixel array used in their 0.9-m telescope imaging camera. Observatories keep lists of each of their instruments and associated CCDs with detailed documentation about the CCD specifications. For examples of such information, check out the observatory websites listed in Appendix B.

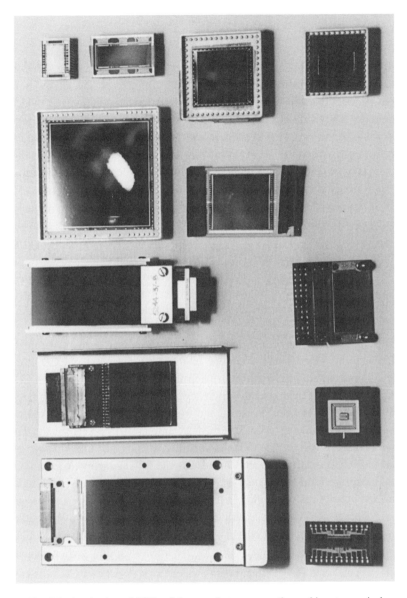

Fig. 1.1. A selection of CCDs of the type that are currently used in astronomical instruments at various telescopes throughout the world. Clockwise from bottom left they are: SITe-002 (2048 × 4096), Loral 2k3eb (2048 × 2048), E2V CCD42-80 (2048 × 4096), SITe-424 (2048 × 2048), GEC P8603 (385 × 578), E2V 15-11 (1024 × 256), TeK1024 (1024 × 1024), Loral 512FT (512 × 1024), E2V-05-30 (1242 × 1152), E2V CCD42-10 (2048 × 512), Loral-64 (64 × 64), and E2V CCD39-01 (80 × 80). E2V Technologies was formerly known as Marconi and prior to that as EEV.

1.2 Why use CCDs?

Most astronomical detectors in use today at professional observatories, as well as with many amateur telescopes, are CCDs. This fact alone gives the reader an impression that there must be something very special or useful about CCDs; otherwise why all the fuss? CCDs have revolutionized modern astronomy. They will take their place in astronomical history along with other important discoveries such as the telescope, photographic plates, prisms, and spectroscopy. The contribution to our knowledge of the heavens brought about by CCDs is astounding, even more so when one considers that they have been in use for only about thirty years.

First introduced as electronic analogs to magnetic bubble memory (Amelio, Tompsett, & Smith, 1970; Boyle & Smith, 1970) at Bell labs, CCDs provided their first astronomical image in 1975 when scientists from the Jet Propulsion Laboratory imaged the planet Uranus at a wavelength of 8900 Å (Janesick & Blouke, 1987). This observation used the University of Arizona 61-inch telescope atop Mt. Lemmon and a prototype version of a CCD made by Texas Instruments Corporation as part of a development project for NASA spacecraft missions.

During the past ten years, tremendous progress has been made in the manufacturing process and therefore in the properties of the CCD itself. These improvements have allowed much lower noise properties for CCDs, thereby increasing their overall efficiency in astronomy. In addition, larger format devices have been produced and the readout times are much shorter, approaching 1-2 seconds even for arrays as large as 1024 pixels square. This latter advance is mainly due to the availability of high-speed, low-power and low-noise CCD controllers (see Chapter 2). The driving technology for CCD manufacturing is for items such as copy machines, TV cameras, and digital cameras, but the requirements for low noise, excellent pixel cosmetics, and nearly perfect performance is still firmly rooted in astronomy. We outline below two of the important reasons why CCDs are considered as essentially the perfect imaging device. Details of the manufacturing techniques and properties of CCDs will be presented in Chapters 2 and 3.

1.2.1 Noise properties

The usefulness of a detector is very often determined by the amount of inherent noise within the device itself. We shall see in Chapter 3 how the noise properties of a CCD are determined, but, suffice it to say here, modern astronomical CCDs are almost noise free. The original line of photosensitive electronic array detectors, such as television-type imaging detectors, vidicons, silicon intensified targets, and image-dissector scanners, all had very high noise properties. For comparison, silicon intensified target imagers (SITs) had a noise level upon readout of 800 electrons per picture element. Some very good systems of this type could be produced with read noise values of only 200 electrons (Eccles, Sim, & Tritton, 1983). The first CCDs had readout noise levels similar to this latter value, while modern CCDs have noise values of ten down to two electrons per pixel per readout. The large noise levels present in early array detectors not only limited the signal-to-noise ratio obtainable for a given measurement, they also severely limited the total dynamic range available to the camera. Another "feature" of older, higher noise CCDs was the decision an astronomer had to make about co-addition of frames. Since the read noise adds as its square to the total noise budget (see Chapters 3 & 4) adding two frames resulted in a much higher read noise contribution. Today, with typical read noise values of 2–5 electrons, co-addition is essentially equal to a single exposure of longer integration time.

1.2.2 Quantum efficiency and band-pass

Quantum efficiency (QE) is the term used to report on the ability of a detector to turn incoming photons into useful output. It is defined as the ratio of incoming photons to those actually detected or stored in the device. A QE of 100% would be an ideal detector with every incoming photon detected and accounted for in the output. Band-pass is a term that means the total spectral range for which a detector is sensitive to the incoming photons. Our eyes, for example, have a very limited band-pass covering only about 2000 Å of the optical spectral range, from near 4500 to 6500 Å.

One of the great advantages of CCDs compared with earlier detectors is their ability to convert a large percentage of incoming photons into photoelectrons. Photographic plates had an intrinsic quantum efficiency of only about 2% (Kodak IIIaJ's reached 3%), with "hypersensitized" plates (plates treated to a black-magic process involving heating and exposure to various "forming" gases) reaching claimed QEs as high as 10%. Because photographic emulsions were originally more sensitive to UV and blue light,[1] numerous dyes and coatings were developed to both extend their band-pass coverage and allow detection of yellow to red optical photons.

---

[1] The fact that early astronomical imagers (i.e., photographic plates) were blue sensitive is a major reason that most of today's standard stars are blue, the MK spectral classification scheme was initially blue feature based, and why astronomical discoveries such as brown dwarfs and high-z quasars did not happen until recently.

Early solid-state imaging devices and intensified silicon target devices could reach quantum efficiencies as high as 20–40%. These devices relied on television-type tube technology and electron beam scanning for readout of detected photons. Since they used silicon (or similar type materials) as the detector material, their useful spectral band-pass was similar to that of modern CCDs. However, besides the relatively low QE, these early electronic detectors had other drawbacks. For example, image tubes needed high voltage input for electron acceleration and the observed two-dimensional scene was not easily or consistently divided into well-determined x, y positional output (Walker, 1987). Changes in the voltage levels, collected charge, and telescope position resulted in electric and magnetic field variations leading to positional and flux measurement uncertainties.

Even the earliest CCDs (those manufactured by Fairchild or GEC) easily reached peak QEs of 40%. Typical CCD QE curves today not only peak near 90% but are 60% or more quantum efficient over two thirds of their total spectral range. Increased red sensitivity using deep depletion techniques and better thinning and coating processes (blue) will be discussed later. The band-pass available in a modern CCD (with a QE of 10% or more) is about 3000–11 000 Å. Coatings and phosphors deposited on the CCD surface or the use of some form of pre-processor device can extend the band-pass sensitivity limits or increase the QE in specific wavelength ranges (see Chapter 2).

This volume of the Cambridge Observing Handbooks for Research Astronomers will explore the world of CCDs from their inner workings to numerous applications in observational astronomy. Appendices are included to provide ancillary information related to the main text. The chapters will make little assumption as to the reader's previous knowledge on the subject, each attempting to be somewhat self-contained. Chapters 4, 5, and 6 deal directly with astronomical applications while Chapters 2 and 3 are of general interest to those wanting an overall understanding of CCDs as detectors. Chapter 7 discusses the use of CCDs at non-optical wavelengths. In a short treatise such as this, coverage of numerous details and nuances is not possible; thus a detailed reference list to general texts or collections of articles on CCDs is provided in Appendix A. For those wishing to explore a subject at a deeper level, pertinent research articles are cited throughout the text itself.

1.3 Exercises

1. Using the manufacturer websites given in Appendix B, make a list of the various CCDs they produce taking note of the physical and pixel sizes

of each. Can you draw any conclusions about a relationship between the CCD size and the application it is designed for?

2. Using two of the astronomical observatory websites listed in Appendix B, make a list of the types of instrumentation available and the specific type of CCD used in each. Can you draw any conclusions about a relationship between the CCD properties and physical size, and the type of instrument and science it is designed for?

3. Read the article mentioned in Chapter 1 in which the first astronomical CCD image is contained. Discuss how this one advance changed astronomical imaging.

4. What are the two most important reasons that CCDs are the detector of choice in modern astronomy? How do these two properties compare between your eye and those of a typical CCD?

2

CCD manufacturing and operation

Before we begin our discussion of the physical and intrinsic characteristics of charge-coupled devices (Chapter 3), we want to spend a brief moment looking into how CCDs are manufactured and some of the basic, important properties of their electrical operation.

The method of storage and information retrieval within a CCD is dependent on the containment and manipulation of electrons (negative charge) and holes (positive charge) produced within the device when exposed to light. The produced photoelectrons are stored in the depletion region of a metal insulator semiconductor (MIS) capacitor, and CCD arrays simply consist of many of these capacitors placed in close proximity. Voltages, which are static during collection, are manipulated during readout in such as way as to cause the stored charges to flow from one capacitor to another, providing the reason for the name of these devices. These charge packets, one for each pixel, are passed through readout electronics that detect and measure each charge in a serial fashion. An estimate of the numerical value of each packet is sent to the next step in this process, which takes the input analog signal and assigns a digital number to be output and stored in computer memory.

Thus, originally designed as a memory storage device, CCDs have swept the market as replacements for video tubes of all kinds owing to their many advantages in weight, power consumption, noise characteristics, linearity, spectral response, and others. We now further explore some of the details glossed over in the above paragraph to provide the reader with a basic knowledge of the tortuous path that the detected photon energy takes from collection to storage. The design of CCD electronics, semiconductor technology, and detailed manufacturing methods are far beyond the level or space constraints of this book. For further information the reader is referred to the excellent discussion in Janesick & Elliott (1992) and Janesick (2001) plus the other technical presentations listed in Appendix A.

2.1 CCD operation

The simplest and very understandable analogy for the operation of a CCD is also one that has been used numerous times for this purpose (Janesick & Blouke, 1987). This is the "water bucket" idea in which buckets represent pixels on the CCD array, and a rainstorm provides the incoming photons (rain drops). Imagine a field covered with buckets aligned neatly in rows and columns throughout the entirety of the area (Figure 2.1). After the rainstorm (CCD integration), each bucket is transferred in turn and metered to determine the amount of water collected. A written record (final CCD image) of the amount of water in each bucket will thus provide a two-dimensional record of the rainfall within the field.

Referring to the actual mechanisms at work within a CCD, we start with the method of charge generation within a pixel: the photoelectric effect.[1] Incoming photons strike the silicon within a pixel and are easily absorbed if

Fig. 2.1. CCDs can be likened to an array of buckets that are placed in a field and collect water during a rainstorm. After the storm, each bucket is moved along conveyor belts until it reaches a metering station. The water collected in each field bucket is then emptied into the metering bucket within which it can be measured. From Janesick & Blouke (1987).

---

[1] Albert Einstein received his Nobel Prize mainly for his work on the photoelectric effect, not, as many think, for relativity.

they possess the correct wavelength (energy). Silicon has a band gap energy of 1.14 electron volts (eV), and so it easily absorbs light of energy 1.1 to 4 eV (11 000 to 3000 Å).[1] Photon absorption causes the silicon to give up a valence electron and move it into the conduction band. Photons of energy 1.1 eV to near 4 or so eV generate single electron–hole pairs, whereas those of higher energy produce multiple pairs (see Section 2.2.8 and Chapter 7). Left to themselves, these conduction band electrons would recombine back into the valence level within approximately 100 microseconds. Silicon has a useful photoelectric effect range of 1.1 to about 10 eV, which covers the near-IR to soft X-ray region (Rieke, 1994). Above and below these limits, the CCD material appears transparent to the incoming photons.

Once electrons have been freed to the conduction band of the silicon, they must be collected and held in place until readout occurs. The details of the actual construction of each pixel within a CCD, that is, the formation of the MIS capacitor with its doped silicon, layers of silicon dioxide, etc., are beyond the scope of this book (Eccles, Sim, & Tritton, 1983; Janesick & Elliott, 1992), but suffice it to say that each pixel has a structure allowing applied voltages to be placed on subpixel sized electrodes called gates. These gate structures provide each pixel with the ability to collect the freed electrons and hold them in a potential well until the end of the exposure. In a typical arrangement, each pixel has associated with it three gates, each of which can be set to a different voltage potential. The voltages are controlled by clock circuits with every third gate connected to the same clock. Figure 2.2 illustrates this clocking scheme for a typical three-phase device.

We note in Figure 2.2 that, when an exposure ends, the clock voltages are manipulated such that the electrons that have been collected and held in each pixel's +10 volt potential well by clock voltage V3 can now be shifted within the device. Note that electrons created anywhere within the pixel during the exposure (where each pixel has a surface area equal to the total area under all three gates) will be forced to migrate toward the deepest potential well. When the exposure is terminated and CCD readout begins, the voltages applied to each gate are cycled (this process is called clocking the device) such that the charge stored within each pixel during the integration is electronically shifted. A simple change in the voltage potentials (V3 goes to +5 volts, while V1 becomes +10 volts and so on) allows the charge to be shifted in a serial fashion along columns from one CCD pixel to another throughout the array. The transfer of the total charge from location to location within the array is not without losses. As we will see, each charge transfer (one

---

[1] The energy of a photon of a given wavelength (in electron volts) is given by $E(eV) = 12407/\lambda(\text{Å})$.

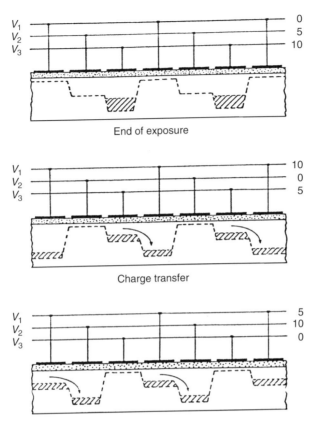

Fig. 2.2. Schematic voltage operation of a typical three-phase CCD. The clock voltages are shown at three times during the readout process, indicating their clock cycle of 0, 10, and 5 volts. One clock cycle causes the stored charge within a pixel to be transferred to its neighboring pixel. CCD readout continues until all the pixels have had their charge transferred completely out of the array and through the A/D converter. From Walker (1987).

of which occurs for each voltage change or clock cycle) has an associated efficiency. This efficiency value is the percent of charge transferred compared with that which was actually collected. Modern values for the charge transfer efficiency (CTE) are approaching 0.999 999 (i.e., 99.9999% efficient) for each transfer.

Each column in the array is connected in parallel and thus each pixel shift is mimicked throughout the entire array simultaneously. One clock cycle moves each row of pixels up one column, with the top row being shifted off the array into what is called the output shift register or horizontal shift register. This register is simply another row of pixels hidden from view (i.e., not exposed to incident light) and serves as the transition between active rows on the array and the output of the device. Once an entire row is shifted into the output register, and before any further row shifts on the active area occur, each pixel in the output register is shifted out one at a time (in a similar manner as before) into the output electronics. Here, the charge collected within each pixel is measured as a voltage and converted into an output digital number (see Section 2.4). Each pixel's collected charge is sensed and amplified by an output amplifier. CCD output amplifiers are designed to have low noise and are built directly into the silicon circuitry; thus they are often referred to as on-chip amplifiers. These amplifiers must work with extremely small voltages and are rated, as to their sensitivity, in volts per electron. Typical values are in the range of 0.5 to 4 microvolts per electron. Figure 2.3 is a microphotograph of an actual CCD showing the various parts we just discussed. In addition, this CCD is an L3CCD (see below) and has an extended serial register the latter half of which is a gain register.

The output voltage from a given pixel is converted to a digital number (DN) and is typically discussed from then on as either counts or ADUs (analog-to-digital units). The amount of voltage needed (i.e., the number of

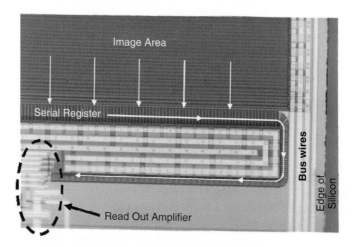

Fig. 2.3. Microphotograph of a E2V L3CCD (see Section 2.2.7) showing the image area (pixels), the serial register, and the on-chip readout amplifier. The other wiring and the bus wires are electrical connections that carry the clock signals and bias voltages to use. Added on to the normal CCD components is an extended serial register through which the readout occurs (the arrow indicates this flow) where the half after the bend is the gain register.

collected electrons or received photons) to produce 1 ADU is termed the gain of the device. We will discuss the gain of a CCD in Chapter 3 and here only mention a few items of interest about it. A typical CCD gain might be 10 electrons/ADU, which means that for every 10 electrons collected within a pixel, the output from that pixel will produce, on average, a count or DN value of 1. For example, with this gain value if a pixel collects 1000 electrons (photons), the output pixel value stored in the computer would be 100 ADUs. For 1500 electrons 150 ADUs would be produced and for 17 234 electrons, the output pixel value would be 1723 ADUs (note, not 1723.4). Digital output values can only be integer numbers and it is clear already that the discrimination between different pixel values can only be as good as the resolution of the gain and digital conversion of the device.

Conversion of the output voltage signal into a DN is performed within a device called an analog-to-digital converter (A/D or ADC). We will see later on that there is an intimate connection between the number of digital bits available in the A/D and the value of the gain that can or should be used for the CCD. The output DNs are usually stored initially in computer memory and then moved to disk for storage and later manipulation.

The process of shifting each entire CCD row into the output register, shifting each pixel along within this register, and finally performing the voltage conversion of each pixel's stored charge by the A/D to produce a DN value is continued until the entire array of pixels has been readout. For large-format CCD arrays, this process can take upwards of a few minutes to complete a single read out of the entire device. Note that for a 2048 × 2048 CCD, the charge collected in the last pixel to be read out has to be transferred over four thousand times. However, most modern large-format CCDs or mosaic cameras containing many large CCDs use a few tricks to readout faster. Single monolithic CCDs usually have 2 or 4 output amplifiers available (one in each corner) and given the proper electronic setup, these large chips are often read out from 2 or 4 corners simultaneously, thus decreasing the total readout time by 2–4. For a mosaic of CCDs, this same process can read the entire array (using multiple amplifiers on each CCD) much faster than even one single large CCD.

The array size of a single CCD, as well as the size of a given pixel on a device, is controlled by the current limitations of manufacturing. How large one can make a good quality, large-scale integrated circuit and how small one can make a MIS capacitor, both of which have demanding requirements for near perfect operation, set the scale of CCD and pixel sizes that are available. CCDs as large as 5040 × 10 080 and 7168 × 9216 pixels and pixels as small as 2–10 microns have been successfully produced.

Modern CCDs have much higher processing standards than even five years ago. Items such as multi-layer registration on the silicon wafer on the photomasks used in the production of the CCD integrated circuit and the ability to make smaller electrical component parts on the wafers (such as output amplifiers) lead to much lower noise characteristics, better pixel charge manipulation, and the ability for faster readout speeds with lower noise. For example, better alignment of the CCD layers in each pixel allow lower clock voltages to be used (as low as 2 volts has been demonstrated) leading to lower overall power consumption. This fact, in turn, allows for items such as plastic packaging instead of ceramic, reducing overall packaging costs, a cost that often rivals that of the CCD itself.

As you might imagine, astronomy is not the driving force for CCD manufacturing. Video devices, cell phones, security cameras, Xerox machines, etc. are the global markets boosting the economy of CCD makers. The trend today is to produce CCDs with small pixels (10–12 microns for astronomy down to ~ 2 microns for other applications) in order to increase image resolution. Small pixels (and small CCDs) have lower cost and higher yield but the small pixels have shallow well depths. This is somewhat compensated for using fast readout techniques and/or microlens arrays, which focus light from an incoming source onto each small CCD pixel. Not all CCD pixels are desired to have shallow wells. The CCDs produced by E2V for the NASA Kepler Discovery mission have 27 micron pixels with well depths of nearly 1 million electrons each and a capacity of $> 50\,000$ electrons per pixel is quite common in astronomy. Even CCDs with built-in electronic shutters are being experimented with. Each pixel contains a p^+-n-p^- vertical overflow drain (VOD) photodiode structure on its top through which the incoming light passes. The absorption of incoming light when the "shutter" is open is minimal and, within a few hundred nanoseconds, the electronic shutter can be biased and become opaque. The interested reader is referred to Janesick & Elliott (1992), Janesick (2001), Kang (2003), Robinson (1988a), and Appendix A for further details.

2.2 CCD types

When reading about CCDs, one of the most confusing issues can be the various terms listed in the literature or in commercial documents. Items such as backside illuminated, buried channel, deep depletion, and antiblooming are just a few. This section will provide a brief discussion to some of these terms while Chapter 3 will discuss CCD characteristics in detail. Further information can be found in the references listed in Appendix A. In particular, readers desiring a microscopic look at the electronic structures of a CCD integrated circuit are referred to Janesick (2001) and the many SPIE articles listed therein. Some terms, such as quantum efficiency and full well capacity, will be used here without proper introduction. This will be rectified in the next chapter.

2.2.1 Surface channel versus buried channel CCDs

As discussed above, charge stored within a pixel is moved from pixel to pixel during readout via changes in electrical potential between the biased gates. For surface channel CCDs, this charge movement occurs "on the surface" of the CCD, being transferred between overlapping gates. For example, each pixel in a three-phase CCD has three overlapping gates for each imaging cell or pixel. Each cell has an associated transfer efficiency and maximum possible transfer rate. The efficiency is near to, but not quite, 100% for each cell (we will discuss this charge-transfer efficiency (CTE) further in Chapter 3). Typical rates of transfer for CCD television cameras (or video cameras) are several megahertz, while low-light level applications with cooled devices (i.e., those occurring in astronomy) use rates nearer to the kilohertz range (Eccles, Sim, & Tritton, 1983).

The major drawback to the surface channel CCD is the presence of trapping states that occur at the boundaries of the gates. These traps, which are caused by imperfections within the silicon lattice or the gate structures, cause a loss of transfer efficiency by trapping some of the electrons from other pixels as they pass by during readout. Traps within CCDs are therefore undesirable for faint light level applications such as astronomical imaging. One method that can be used to eliminate traps (although not in general use anymore because of the advent of buried channel devices; see below) is to raise the entire surface charge level of the CCD above the level needed to fill in any traps. This is accomplished by illuminating the CCD at a low level prior to exposure of the astronomical source. This technique, called a pre-flash or a fat zero, allows any nonlinearities resulting from traps at low light levels to be avoided while only slightly increasing the overall noise level of the resulting image (Djorgovski, 1984; Tyson & Seitzer, 1988).

Deferred charge is another source of nonlinearity sometimes present at low light levels (Baum, Thomsen, & Kreidl, 1981; Gilliland, 1992). Often referred to as charge skimming, charge depletion, or low light level nonlinearity, this particular worry has all but disappeared in modern, low-readout-noise devices. We will address nonlinearity in CCDs in a more general manner in Section 3.8.

A better solution than those just discussed is to move the charge from pixel to pixel during readout via a channel of semiconductor material that is away from the gate structures. This "buried channel" architecture results from the application of an additional layer of semiconductor material placed under the CCD surface. The buried channel greatly enhances the charge movement through the device by reducing the number of possible trap sites and by decreasing the transfer time between adjacent cells. Higher transfer rates (up to 100 MHz) and high CTE values (>99.995%) are easily accomplished with buried channel devices.

The price paid for the addition of the buried channel is that the total charge storage capacity for each pixel is reduced by 3 or 4 times that of a surface channel detector. However, since the operation at very low signal levels is much improved, the overall dynamic range and sensitivity of a buried channel device become much higher.

2.2.2 Front-side and back-side illuminated CCDs

CCDs are manufactured as single large-scale integrated devices. They have a front-side, where all the gate structures and surface channel layers are deposited, and a back-side, which is simply bulk silicon generally covered with a thin conductive layer of gold. CCDs used as front-side illuminated devices work as the name implies, that is, illumination occurs on the front of the CCD with the photons being absorbed by the silicon after passing directly through the surface gate structures. The device thickness is of order 300 microns from front to back making these chips relatively high in their susceptibility to detection of cosmic rays. Because the photons must first pass through the gate structures before they can be absorbed by the silicon, front-side illuminated CCDs have lower overall quantum efficiencies than the back-side devices (discussed below). However, front-side devices provide a flatter imaging surface and the actual CCD itself is easier to handle and work with. Figure 2.4 provides a schematic view of a single front-side illuminated pixel.

Back-side illuminated devices, also known as thinned devices, are again justly named. The CCD, after manufacture, is physically thinned to $\gtrsim 15$ microns by various etching techniques (Lesser, 1994). The device is then mounted on a rigid substrate upside down and illuminated from behind. The incoming photons are now able to be absorbed directly into the bulk silicon pixels without the interference of the gate structures. The advantages in this type of CCD are that the relative quantum efficiency greatly exceeds that of a front-side device and the response of the detector to shorter wavelength

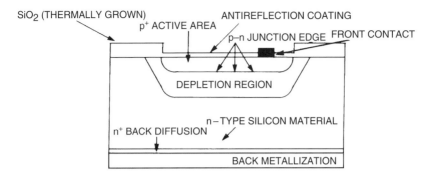

Fig. 2.4. Schematic view of a single front-side illuminated CCD pixel. The square labeled "front contact" is a representation of part of the overall gate structure. The letters "p" and "n" refer to regions within the pixel consisting of silicon doped with phosphorus and boron respectively.

light is improved since the photons no longer need to pass through the pixel gates. Disadvantages of back-side devices are in the areas of shallower pixel well depths (due to the much smaller amount of material present per pixel), possible nonuniform thinning leading to surface and flat-field nonuniformities, and increased expense incurred by the thinning and mounting process.

In the next chapter, we will explore the idea of quantum efficiency further and provide an example of the large differences present in these two types of CCDs. Back-side devices generally have about twice the quantum efficiency for a given wavelength compared with front-side devices.

2.2.3 Interline and frame transfer CCDs

Interline transfer CCDs are specially constructed devices in which each column of imaging (active) pixels is paralleled by a light-shielded column of storage (or inactive) pixels. The device is used as a normal CCD but after the exposure ends, each light-sensitive column of data is quickly shifted into its neighboring light-shielded column. The shift occurs in only a few microseconds and so image smear is almost nonexistent and the need for a mechanical shutter is precluded. The shifted image data can then be clocked out of the device while the active columns are again integrating on the source of interest. Interline devices have been used in many of the world's high-speed photon counting array detectors owing to their fast shift operation (Timothy, 1988).

The imaging area in an interline device is not continuous as each active column is paralleled by an inactive one; thus there is an immediate reduction in the overall extrinsic quantum efficiency by a factor of two. The electrode and gate structures of the active pixels are a bit different from those in a normal CCD, causing a further reduction in quantum efficiency and a somewhat different overall spectral response. Due to these two factors and the fast readout times of modern CCDs, interline devices are generally not seen in astronomy much anymore.

Frame transfer devices work in a manner similar to that of having two separate CCDs butted together. One half of the device is active (exposed to incoming light) and records the image while the other half is shielded and used as a storage device. The end of the integration sees the image data quickly transferred or shifted to the identical storage half, where it can be readout slowly as a new integration proceeds on the active side. Frame transfer CCDs are used in most commercial video and television cameras for which the active image is readout at video rates (30 frames per second). Astronomical imaging generally can not occur at such high rates because of photon starvation, but frame transfer devices have been used in a number of interesting astronomical instruments. For example, ULTRA-CAM at the 4.2-m William Herschel telescope (Dhillon & Marsh, 2001) and a CCD time series photometer at the University of Texas (Nather & Mukadam, 2004). More on high-speed CCD observations will be presented in Chapter 5.

Figure 2.5 provides an illustration of these two types of CCD. We note here that in both of these special purpose CCD types, the movement of the accumulated charge within each active area to the shielded, inactive area can be accomplished at very high rates. These rates are much higher than standard CCD readout rates because no sampling of the pixel charge or A/D conversion to an output number occurs until the final readout from the inactive areas. As we will see, the A/D process takes a finite time and is one cause of noise added to the output signal. Therefore, any on-chip shifting or summing of charge will introduce essentially no additional noise into the data.

Before we leave this section, we are compelled to discuss a non-astronomical use of CCDs that is likely to be of interest to some readers of this book. Megapixel multiple-frame interline transfer CCDs (MFIT CCDs) produced with alternating sets of three pixels covered with red, green, and blue filters and which are readout at 30 frames/second have become common. Have not heard of them you say? Well, these are the imaging devices used to bring you high definition TV (HDTV). So the next time you watch an HDTV program (this author has yet to see one) sit back and smile realizing that the event is brought to you by your old friend, the CCD.

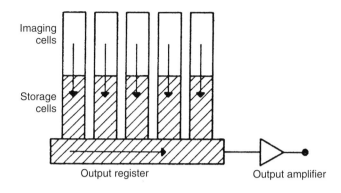

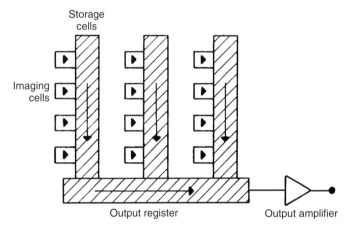

Fig. 2.5. Cartoon view of (top) a frame transfer CCD and (bottom) an interline CCD. From Eccles, Sim, & Tritton (1983).

2.2.4 Antiblooming CCDs

Observations of bright objects can lead to the collection of too many photoelectrons within one or more of the pixels of a CCD, causing the effected pixels to overflow.[1] This condition will occur when the full well capacity of a pixel (see next chapter) is exceeded. The result of such saturation is termed bleeding and is easily noted on a CCD image as one or more bright pixels with a bright streak leading away from the region, generally in two opposite

---

[1] Think here of the buckets in the field during a rainstorm and what would happen if too much water fell into any given one of them.

directions.[1] Saturation effects can be irritating or, even worse, the streak can bleed into an object of interest rendering its measurement useless. Orthogonal transfer CCDs (see below) have a different pixel structure and thus do not show bleeding. Instead, saturated objects produce a puddle on an OTCCD. Figure 2.6 shows a CCD image with a saturated star.

To counter pixel saturation behavior, one could simply integrate for a shorter period of time. This will cause the offending object to supply fewer photons per integration and thus not to saturate any CCD pixels. However, to obtain the same exposure level, a number of frames would have to be taken and co-added together later. We will see in the next chapter that this is not always equivalent (in noise terms) to a single long exposure. Therefore, although shorter exposures are a simple approach, they are not always practical. Many times we wish to study a field in which objects of various brightness are interspersed, presenting the need for considerable dynamic range within a single image. Thus, the antiblooming gate was invented (Neely & Janesick, 1993).

Antiblooming gates have added electrical structures manufactured into the CCD itself, allowing saturated pixels to be "drained" off and controlled instead of bleeding into neighboring pixels, destroying their integrity. The offending pixels, in such a case, are still not usable, but the neighboring pixels are. The trade-off that occurs for a CCD with antiblooming gates is the following: about 30% of the active pixel area is lost (as it becomes the drain gate) giving a remaining active pixel area of only 70%. The use of such a gate structure is therefore equivalent to an effective QE reduction. Such additional gate areas also reduce the overall spatial resolution of any measurement (as the drain gates leave a gap between adjacent pixels) and lower the total pixel well depth by about one half. A rule of thumb here is that a typical antiblooming gate CCD requires about twice the integration time to reach the same signal level as a similar device without antiblooming gates.

Antiblooming gates are currently available on certain makes of CCD. When deciding if they are right for you, keep the following in mind: Longer exposures will be required for a given signal level, and if working in a spectral

---

[1] The reader should be aware of the fact that CCDs can saturate in two different ways. As discussed here, saturation can occur if a given pixel is exposed to too much light and overflows, exceeding its full well capacity. This type of saturation leads to the result discussed here (see Figure 2.6), that is, bleeding of excess charge into neighboring pixels. The other way in which a CCD can saturate is if any given pixel collects more charge than can be represented by the output from the A/D converter (see Sections 2.4 and 3.8), even if the amount collected is below the full well capacity of the pixel. This type of saturation does not lead to charge bleeding but yields such oddities as flat-topped star profiles or constant (exact) pixel values over the region of A/D saturation. As the amount of charge collected by the pixels within a CCD increases, eventually one of these two types of saturation will set the usable upper limit for a given system. We will see that the linearity of a CCD (Sections 3.6 and 3.8) may also limit the maximum usable ADU level of a given observation.

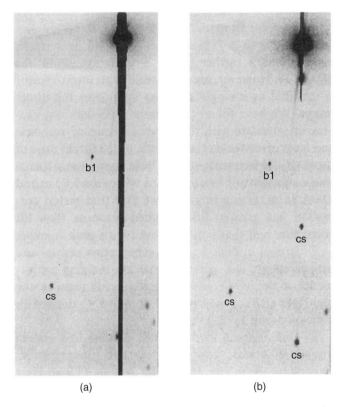

Fig. 2.6. Two equal-length CCD exposures of a bright star (SAO 110456). The normal CCD exposure (a) shows typical bleeding caused by saturation within the CCD. The CCD exposure on the right (b) was made with an antiblooming CCD and clearly shows the much reduced bleeding from the bright star. From Neely & Janesick (1993).

region of already low or reduced quantum efficiency, antiblooming gates will reduce the signal even further for a given integration time. On the plus side, if bleeding of bright objects will be a detriment to your science, the antiblooming gate option may be a good choice.

2.2.5 Multipinned phase (MPP) CCDs

A method developed to achieve ultra-low dark current in a CCD is that of multipinned phase operation or MPP mode. CCDs operating in MPP mode are common in inexpensive, "off-the-shelf" devices for which thermoelectric cooling techniques are exploited. The MPP design relies on the inversion of all three clock voltages in connection with additional semiconductor doping (usually boron or phosphorus) of the phase three gate. The effect of this semiconductor architecture is to greatly lower the dark (thermal) current and this can even allow room temperature operation if the light levels are sufficient. Generally, astronomical low light level applications require CCDs to be cooled to near $-100°$ C before usable images can be obtained. Dark currents in MPP mode CCDs are near 0.01 electrons per second per pixel at temperatures of only -40 to $-65°$ C, levels that are equal to non-MPP mode operation at $-100°$ C.

MPP mode CCDs have allowed a new revolution in CCD technology. Items such as digital cameras (CCDs operating at room temperature) are now commonplace. The lower thermal noise, even at room temperature, allows integration and readout (for high light level scenes) to be used to produce digital photographs. While a boon to higher temperature operation, the full well depth of each pixel is reduced by two to three times that of the same device in non-MPP mode. The higher temperature operation must be balanced against the much reduced dynamic range of the modified CCD. New design strategies and higher quality silicon wafer technology are making progress in restoring the full well depth while keeping the advantages of MPP operation. A CCD modified to run in MPP mode has altered electronic design and pixel structures and, therefore, often cannot be made to operate normally. Thus, an MPP CCD does not allow one to decide which mode of operation to use nor does it provide the ability to switch between MPP and normal operation.

2.2.6 Orthogonal transfer CCDs

We have seen that a typical CCD has three gates per pixel and that the readout operation is via clocking the pixel charge in a three-phase fashion. The charge within a given pixel can only be shifted in the vertical direction, that is, from row to row, until it reaches the output register. This is due to channel stops running vertically, which are biased to keep electrons in a given column. One sees a dramatic example of electron trapping along columns in Figure 2.6. A new type of CCD, the orthogonal transfer CCD (OTCCD), has been developed that allows each pixel's charge to be shifted both vertically and horizontally on the array (Burke *et al.*, 1994).

The ability to move the charge in both directions within the OTCCD comes about through the use of a complex, four-phase mode of operation in which the channel stops become an addition gate. Four gates are used in each pixel, two of which are triangular and make up the central part and two of which are rectangular but smaller in area and are split into pairs surrounding the pixel center. The larger number and size of the gates lead to a lower overall QE for the early OTCCDs and more detailed intra-pixel QE effects are to be expected (Jorden, Deltorn, & Oates, 1993; Jorden, Deltorn, & Oates, 1994).

The first application of an OTCCD was to compensate for image motion during an integration, similar to a low order adaptive optics tip-tilt correction (Tonry, Burke, & Schechter, 1997). A 1024×512 OTCCD device was built to allow one half of the OTCCD to image, quickly readout, and centroid on a bright star, while the other half was used to integrate on a scene of interest. As the bright star center wandered during the integration, the object frame half of the OTCCD was electronically shifted (both vertically and horizontally) many thousand times per 100 seconds in order to move the image slightly (~ 0.5 arcsec) and follow the bright star centroid position. The final result from the OTCCD was an image with much improved seeing and little loss of information.

The complex gate structure of the original OT design caused small dead spots within each pixel as well as allowing charge to be trapped in numerous pockets within a pixel during charge shifting. These two effects amount to about 3% flux losses, which were probably compensated for in the improved final image. Advances in gate structure design and impurity control in the device composition have eliminated most of these losses (Tonry, Burke, & Schechter, 1997, Burke *et al.*, 2004). The OTCCD promises to be extremely useful for certain applications in astronomical imaging.

A next generation OTCCD camera, dubbed OPTIC, the Orthogonal Parallel Transfer Imaging Camera (Tonry *et al.*, 2004) contains two 2K by 4K OTCCDs. Two end regions on each OTCCD (of size ~ 2048 by 512 pixels) are used for guide stars. The ideal operation of OPTIC uses four guide stars, one in each quadrant of the large format CCDs to provide real time, "no moving parts" tip-tilt corrections to the science regions. All the problems with pockets and dead cells in the OTCCDs have been eliminated. Howell *et al.* (2003) present a number of observational extensions to the original purpose of the OTCCD, using it for moving object charge tracking, high-speed, high-precision photometry, and actual shaping of the point-spread function during integration. More on the photometric results available from OTCCDs will be presented later.

OPTIC has performed so well that two large imaging projects (the WIYN observatory one-degree imager and the Pan-STARRS cameras) are planning to use a new generation of OTCCD in their focal plane. Figure 2.7 shows the schematic design for the new style of OTCCD. The individual OTCCDs (approximately 512×512, 10–12 micron pixels) are arranged in a single monolithic 8×8 checkerboard pattern called an orthogonal transfer array

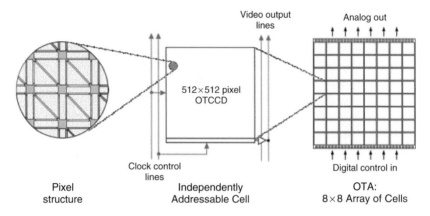

Fig. 2.7. Cartoon of the pixel and CCD layout of the new generation of orthogonal transfer CCDs. In this schematic, we see how the OT pixels are arranged in small ($\sim 512 \times 512$) devices (to increase yield and provide local guide stars) and these individual OTCCDs are placed into a square array of 64 independent devices called an orthogonal transfer array (OTA).

(OTA). Each OTCCD in the OTA is independently controllable and can be used for science imaging, guide stars, fast readout, or simply turned off (Burke *et al.*, 2004).

Figure 2.8 shows one of the first production silicon wafers made by STA/Dalsa. The wafer contains three OTAs, a few other CCDs, and some test devices. This wafer represents one of the first OTA wafers to be processed for the new WIYN observatory imagers.

2.2.7 Low light level CCDs - L3CCDs

A low light level CCD (L3CCD) is a completely normal CCD with an extended serial or output register added in to the design. The extended section of the serial register, the gain register, is clocked out with a higher voltage than normal (40–60 volts) allowing for a slight possibility (1–2%) that each electron transferred through each gate will create a second electron by avalanche multiplication. The extended gain (or multiplication) register thus has a total probability of a gain equal to 1.01^N, where N is the number of elements in the gain register. For a typical N of ~ 500, the L3CCD produces an on-ship gain of 145.

The electrons finally pass through a normal style CCD output amplifier and are digitized. Read noise (Chapter 3) is the noise in this output electronic stage and the equivalent read noise from an L3CCD is the read noise reduced by the

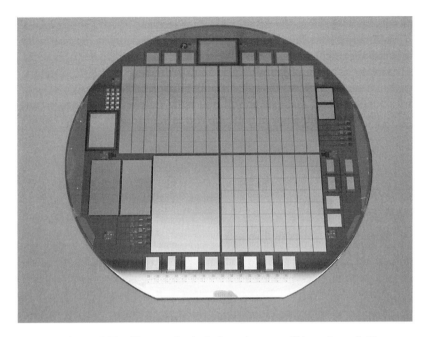

Fig. 2.8. An OTA silicon wafer fresh from the oven. Thin wafers of silicon are used along with various chemical processes and photomasks to produce the integrated circuits we call CCDs. Here we see a 150 ohm-cm Si wafer onto which have been built three 12 micron pixel OTAs (the checkerboard devices), two 2.6 K by 4 K, 12 micron pixel standard CCDs, two 2K by 1K CCDs, an 800 by 800 CCD, and a number of other smaller test CCD devices. The standard CCDs were used to determine the validity of the wafer production as well as being employed in a variety of instruments at the WIYN observatory, NOAO, and elsewhere. The test devices were used to provide feedback on a variety of electronic design and configuration tests used in the OTAs. The CCDs will be cut apart and mounted separately for use. The Si wafer is 15 cm in diameter.

gain factor. Thus for low noise, slow readout CCDs, L3CCDs are often called noise-free. Even at very fast readout rates (10–100 frames/second), which we will see can greatly increase the read noise, they are quite respectable. L3CCDs are still a very new product being developed for security cameras, night vision applications, and fast readout requirements. To date they come in small formats (1000 × 1000 pixels with most being much less) of pixel sizes from 20–40 microns (E2V L3CCDs) to a version with 7.4 micron pixels (TI L3CCD). The number of gain stages, N, is near 400–600 and some astronomical applications have been tested.

The ESA space mission GAIA will use L3CCDs in one of its on-board CCD cameras. L3CCDs work as nearly perfect photon counting devices and promise to be a great addition to astronomical imagers. See Mackay *et al.*, (2002) and Jerram *et al.*, (2001) for more information. Appendix B lists some websites with additional information on L3CCDs.

2.2.8 Superconducting tunnel junctions

Although not truly a CCD, the superconducting tunnel junction (STJ) is a recent semiconductor device addition to the field of astronomical imaging. Using a layer of polycrystalline niobium (Nb) supercooled to a temperature of only a few hundred millikelvin, an STJ diode acts as a photon detector (Perryman *et al.*, 1994). Used initially as an X-ray photon counter, the STJ is seeing new applications being developed in the areas of UV, visible, and IR light detection.

In Chapter 7, we will discuss how CCD detection of a high energy photon is accomplished and additionally how this detection and its associated photoelectron production provide information on the energy (wavelength) of the incoming light particle.[1] In general, light of low energy (UV to IR) does not produce multiple free electron pairs within a semiconductor; thus each incident photon (regardless of wavelength) looks like all the rest upon detection. Separation of spectral bands must be accomplished by filtering the input beam.

Supercooled Nb STJ devices, however, allow UV, optical, and even IR photons to produce numerous pairs of free electrons, thus allowing an approximate determination of the energy (wavelength) of each incoming particle as well as producing a normal image. Initial application of an STJ device (Peacock *et al.*, 1996) has shown sensitivity to wavelengths of 2000 to 10 000 Å with a QE of near 50%. Spectral resolution of the incoming light was about 450 Å, but it was hoped that future devices made of other superconducting materials, such as aluminum, might provide wavelength coverage from 10 Å to well over 1 micron at spectral resolutions of 80 Å.

The first astronomical test of an STJ occurred during the fall of 1998 at the Observatorio del Roque de los Muchachos on La Palma in the Canary Islands. A single STJ diode, placed at the telescope focus, produced a broadband spectrum of a star by simply imaging it (Perryman *et al.*, 1999). Since that time, tantalum strip absorber Al STJ devices have appeared and have been placed in service at the 4.2-m William Herschel telescope. This camera, S-Cam-2, consists of a 6×6 array of 25 micron square Ta junctions, which record photon arrival times to 5 microseconds and provide a spectral resolving

---

[1] In the world of photon counting astronomy, using the photoelectrons produced to determine the wavelength (energy) of each incident photon is termed pulse-height spectroscopy.

power of R ∼ 8 or about 600 Å in the optical region. A discussion of this
camera and one application is presented in Reynolds *et al.*, (2003). Plans to
produce a larger array (10×12 junctions, doubling the field of view) are
underway (Martin *et al.*, 2003) as well as a design for a high resolving power
STJ spectroscope (Cropper *et al.*, 2003). Pure Al junctions, operated at 50 mK
in an adiabatic demagnetization refrigerator, and contacted via Nb electrodes,
have been fabricated into arrays with initial lab and on-sky tests performed
(Shiki *et al.*, 2004). The energy resolution appears to be about twice as good
as previous devices but further testing is needed.

While superconducting tunnel junctions are a far way off from being a
routine detector in use at an observatory, prototype cameras do exist and
extensive testing is underway. After all, it took CCDs nearly ten years after
their introduction to become commonplace at observatories world-wide.

2.2.9 CMOS devices

CMOS detectors or Complementary Metal Oxide Semiconductor arrays (also
called active pixel sensors, APS) are becoming the imager of choice in many
commercial applications. Digital cameras and digital video (TV) cameras are
mostly CMOS arrays and many other applications use them as well.

These devices are nearly identical to CCDs but they contain additional
circuitry built into every pixel. Digital logic for pixel addressing, readout,
and A/D conversion are produced as a part of the array. More complex
devices even provide on-chip signal processing allowing true digital output
to the host computer.

Due to their construction methods, the active photosensitive region per pixel
is very small, thus their overall quantum efficiency is low, being 10–20% in
some devices but up to 35–40% in recent versions at 5500 Å, falling to 15%
at 9000 Å.

The low QE and the higher production cost as well as the lack of maturity
for CMOS devices at present are the main reasons they are not being used
in astronomy. However, back-illuminated CMOS devices and newer pixel
designs have been tested and are on their way to rivaling conventional CCDs
as astronomical imagers. We all could appreciate a detector that would not
only collect the photons but signal process them on the chip and output the
reduced digital values to our awaiting hard drives.

To learn more about CMOS devices see Janesick *et al.* (2002, 2004),
Kang (2003), Woodhouse (2004), and references in Appendix A. Most of the
information on these devices is non-astronomical based at present and mostly

contained in SPIE type articles. However, Bonanno *et al.* (2003) discuss their use of a new generation CMOS-APS device for astronomical purposes.

2.3 CCD coatings

To make use of the properties of CCDs in wavebands for which silicon is not sensitive or to enhance performance at specific wavelengths, CCDs can be and have been coated with various materials.[1] These coating materials allow CCDs to become sensitive to photons normally too blue to allow absorption by the silicon. They generally consist of organic phosphors that down-convert incident UV light into longer wavelength photons, easily detected by the CCD. These vacuum deposited coatings are often used with front-side illuminated, thick CCDs (although not always; the Hubble Space Telescope WF/PC I CCDs, which were thinned devices, were phosphor coated) and can be viewed as an inexpensive method of increasing blue response without the cost and complexity of thinning the device.

A coronene phosphor has been commonly used in recent years to convert photons shortward of 3900 Å to photons at a wavelength near 5200 Å. The use of such a coating causes a "notch" in the overall QE curve as the CCD QE is falling rapidly near 4000 Å but coronene does not become responsive until 3900 Å. Another common phosphor coating, lumogen, eliminates this QE notch, as it is responsive to wavelengths between 500 and 4200 Å (see Figure 2.9). An interesting side note is that lumogen is the commercial phosphorescent dye used in yellow highlighting pens. The QE of a coated CCD may increase in the UV and shortward regions by amounts upwards of 15–20%. Transparent in the visible and near-IR regions, properly deposited coatings can also act as antireflection (AR) coatings for a CCD. More on the use of coated CCDs is contained in Chapter 7.

The main problem encountered when using coated CCDs is that the deposited material tends to evaporate off the CCD when under vacuum conditions (see Chapter 7). A less important problem is the loss of spatial resolution due to the finite emission cone of the light generated in the coating. The specifics of coating materials and the process used to coat CCDs are both beyond the scope of this book. Appendix A and Janesick & Elliott (1992), Lesser (1990), Schaeffer *et al.* (1990), and Schempp (1990) provide further readings in this area.

---

[1] Non-optical operation of a CCD can also be accomplished in this manner. See Chapter 7.

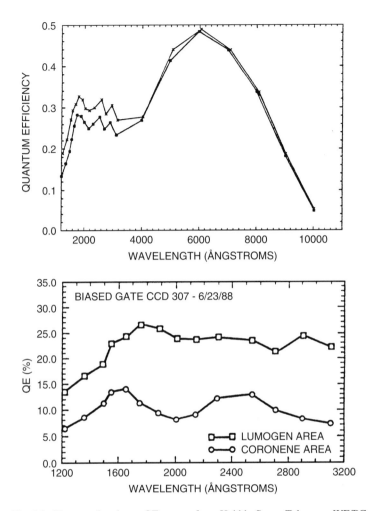

Fig. 2.9. The top plot shows QE curves for a Hubble Space Telescope WF/PC prototype CCD before and after being coated with lumogen. Note the increased UV response of the coated CCD. The bottom plot shows the QE properties of a WF/PC prototype in the far-UV spectral region. Presented are two curves, one for a coronene coated CCD and one for a lumogen coated CCD. From Trauger (1990).

Current processing techniques have reached a state where a given CCD can be "tuned," via its make-up, resistivity, thickness, and operating temperature, to provide a desired response at a specific wavelength (Lesser, 2004). Basically, one can build a pixel structure, QE, and well depth to suit a specific need. Much of the ability to tune CCDs is limited to the red end of the optical spectrum. These topics will be discussed in the next chapter.

2.4 Analog-to-digital converters

Analog-to-digital (A/D) converters are not really a subject for this book. However, the output pixel values from a CCD (digital numbers stored in a computer) are completely determined by the method used to sample the charge on a pixel and how that sampling is converted into a digital number or data number to be stored in computer memory. A/D converters for use in CCDs, including linear and logarithmic types, have been reviewed in Opal (1988). The output from each pixel in a CCD must be examined by electronic means in order to determine how much charge has been collected during a given integration and to convert that charge value into a digital number.

As each packet is clocked out of the CCD output register, it is passed through an on-chip amplifier circuit built directly into the CCD substrate. Next, the packet passes into an external low-noise solid state amplifier, to increase the signal strength, followed by input into the integrating amplifier. The output from the integrating amplifier is then sent directly into an A/D converter, which converts the analog voltage into a digital number to be stored in computer memory. The electronic circuitry just described, minus the on-chip amplifier, is often called the CCD head electronics and is usually located in a box very near or attached directly to the CCD dewar itself.

To measure the voltage of each charge packet, in preparation for conversion into a digital number, a difference measurement is performed between the reset voltage and the sum of the charge packet plus a constant reset voltage. A well-regulated constant voltage source is needed in this step to supply the voltage to the electronics of the inverting plus integrating amplifier circuits. An integrating capacitor, connected across the integrating amplifier, first samples the reset voltage passed through the inverting amplifier for typically $20\,\mu s$. A single pixel charge packet is then passed through the non-inverting amplifier with the output being sampled by the same integrating capacitor for the same time period. Equal time sampling is critical as one has to worry about charge decay in the capacitor. Generally, however, the RC time constant for the capacitor is of the order of a few seconds, while the sampling is a few tens of microseconds. Both signals sampled by the capacitor contain the reset voltage, but only one has the pixel charge packet. Thus the difference formed by the two opposite signed voltages provides a very good estimate of the actual charge level within the sampled pixel. The value of the reset voltage, whatever it may be, is completely eliminated by this differencing. This technique, for sampling the amount of charge within each pixel, is used in almost all current day CCDs and is called correlated double sampling (CDS). Details of the CDS process and some CDS variations such as the dual slope process and triple correlated sampling are discussed in Janesick & Elliott (1992), Opal (1988), Joyce (1992), and McLean (1997b). CMOS devices, as discussed above, can use logic built directly onto each pixel to perform this conversion and provide a digital output directly from the CMOS device.

The assignment of an output digital number to the value of the charge packet from each pixel is the job of the A/D converter. The charge stored in each pixel has an analog value (discrete only at the 1 electron level[1]) and the process of CDS decides how to assign each pixel's charge packet to a specific data number (DN) or analog-to-digital unit (ADU) (Merline & Howell, 1995). As briefly mentioned in Chapter 1, this assignment is dependent on the device gain as follows.

If the gain for a CCD has a value of 10 electrons/ADU, the digitization process tries to divide each pixel's charge packet into units of 10 electrons each, assigning 1 ADU to each 10 electrons it measures. If, after division, there are fewer than 10 electrons left then the remaining charge is not counted and is lost and therefore unknown to the output pixel value. We will see (Section 3.8 and Howell & Merline (1991)) that the gain of a device can have a great effect on its ability to provide good photometric information.

The ultimate readout speed of a given CCD depends on how fast the process of pixel examination and A/D conversion can take place. Modern large-format devices can, and often do, use two or four output amplifiers during readout (one at each device corner) providing a faster overall readout time. However, such readout schemes can introduce systematic offsets between each quadrant of the CCD owing to slight differences in the four output amplifiers and associated electronics. This effect is often seen in images produced by large-format CCDs, where one or more quadrants will show an apparently different background level. An additional common effect seen in many large format CCDs and mosaic CCD imagers is A/D ghosting. If a single CCD or two CCDs in a mosaic share an A/D converter, a bright (usually saturated) star imaged in one part of a CCD (or one CCD in a mosaic) can produce a low level, ghost image in the other CCD region sharing the A/D. This ghosting occurs as the saturated star's charge overwhelms the capacitor during the reset

---

[1] Even though CCDs are called digital devices, the output from each pixel is really an analog voltage that must be measured and digitized by the A/D converter.

cycle and leaves residual charge. This charge is added to the pixels from the shared region producing the ghost image.

The readout speed of a CCD is also related to the number of bits of precision desired (A/D converters with fewer bits of precision work faster than those with more bits) and the acceptable readout noise level (faster pixel readout increases the introduced noise level). Current ultra-low-noise A/D converters can distinguish the reference value from the collected charge in a given pixel at the 2 to 4 electron level.

Let us examine A/D converters in a bit more detail. As CCDs become better produced integrated circuits with less noise and better charge clocking and on-chip parts such as the output amplifiers are made smaller (less thermal noise), the overall noise of a device becomes very low, perhaps 1-2 electrons. At this level, very good modern systems are discovering that the linearity of the A/D converter is now of increased concern. A 1% linearity value is often quoted for CCDs and often observers expect no less from a CCD. But what is the reality of the A/D linearity for a system one is likely to use?

Readout speed has increased, as mentioned above, with an additional reason being the size of the A/D converter and the speed at which it can, well, convert. This entire process is called digitization. CCD controller technology in the 1990s, such as a Leach (SDSU) controller, readout CCDs at a rate of 200 kHz while today's modern controllers A/D convert at rates as high as 800 kHz to 1 MHz. Additionally, present day A/D converters of high quality can be obtained that use 18 bits, not a mere 16 bits. We will see in the next chapter that this is of great benefit.

To complete the efficiency and speed we desire in modern CCD systems, CCD controllers (i.e., the hardware that controls the electronic functions of a CCD) must be quite complex in design, highly versatile in their abilities, and yet simple in their use. Readout rates for CCDs are a prime concern as larger format chips may require excessively long readout times. Controllers such as those developed by San Diego State University (Leach, 1995) can read pixels out at a maximum rate of 10 μs/pixel, but a practical rate of near 50 μs/pixel is generally used to keep the read noise to a minimum (<10 electrons/pixel/read). Even at these seemingly fast rates (20 kHz), a 2048×2048 CCD, containing over four million pixels, takes over 200 seconds to readout the full frame. Requirements for new generation CCDs are even more stringent. A modern CCD controller must readout fast while keeping the read noise below 5 electrons. Complex readout schemes, real-time feedback modes, and other CCD observing strategies mean that modern controllers must also provide the ability to support various observing modes and run various types of CCDs (often including IR arrays). Most controllers under

Fig. 2.10. A photograph of a modern CCD controller. The MONSOON CCD controller is being developed at the NOAO for use with optical CCDs, OTCCDs, and IR arrays. The photograph shows the three component circuit boards and the chassis into which the boards are placed. The three boards are (left to right) a clock and bias board, a master control board, which produces the clocking for integration and readout, and an 8-channel CCD acquisition board, which contains the A/Ds.

development at present have the ability to "plug in" a CPU directly on the controller board and off-load many of the CCD control and readout functions from the observing computer. This allows on-board logic for fast control and readout while keeping the higher level user computer free for real-time data examination and analysis. An example of a modern CCD controller, MONSOON, is shown in Figure 2.10.

2.5 Exercises

1. Make an arrangement of water buckets as shown in Figure 2.1. Have one person use some method of producing an artificial rainstorm over the buckets. Measuring the volume of each bucket, as would be done for the charge collected in each pixel in a CCD, make a plot of the amount of water

in each bucket and attempt to determine the two-dimensional nature of the rainstorm. Try various types of rainstorms and see how well your "readout" technique determines the method by which the buckets were filled.

2. Calculate the energy of photons, in eV and in watts, for wavelengths of 100, 1000, 10 000, and 100 000 angstroms. How do these energies compare with each other? How do they compare to a 100-watt light bulb? How many 5000 Å photons would you need to make the equivalent of a 100-watt light bulb?

3. Convince yourself, by drawing a diagram similar to that in Figure 2.2, that the three-phase operation of a CCD will actually shift charge along a column. Can you design a two-phase CCD?

4. Assuming you are using a modern day CCD with a charge transfer efficiency of 0.99 (i.e., 99.0%), determine how much charge you will lose from the last pixel readout in a 2048 × 2048 array. Determine what CTE is needed in order to produce an acceptable loss of charge for this same array.

5. Describe the difference between a row and a column in a CCD. Does charge shift along rows or along columns? Draw a picture of part of a CCD and label the rows and columns. Draw an arrow in the direction of charge readout starting in a given pixel, passing through the output register, and ending in the A/D converter.

6. If the gain of a specific device is 4.5, what will be the output data number from the CCD if the actual charge collected in a given pixel is 450 electrons? 670 electrons? 20 002 electrons? What happens to the fractional part of the output numbers?

7. Perform an experiment with your own analog-to-digital converter. Using a voltage meter and a variable voltage source, have one person slowly, but in an unknown manner, regulate the voltage. Every 30 seconds over a 5 minute period, read the meter voltage and write down its value rounded to an integer. Make a plot of your values as a function of time. Do you think reading the meter every 30 seconds was accurate enough? Do you think your rounding was accurate? How might you improve the experiment regarding the collected data?

8. Assuming that a CCD pixel is a uniformly made cube of silicon, calculate and plot the relation between the volume of the cube and its surface area for five different pixel sizes. This relation between pixel size and full well capacity is roughly true for real CCDs. Discuss what applications might require a CCD with large pixels and with small pixels. If a CCD has pixels that hold only a small amount of charge while an application requires collection of large amounts of charge, how might this be compensated for in the method one uses to make the observations?

9. Discuss how a "fat zero" or pre-flash helps in eliminating trapped charge in a surface channel CCD.

10. Figure 2.4 illustrates a typical front-side illuminated CCD pixel. Make an approximate drawing of a typical back-side illuminated CCD pixel. Which of these are called "thinned" devices? Why? Discuss some observational applications that would benefit from a front-side illuminated CCD.

11. Discuss why it is that an interline CCD starts with only one-half the quantum efficiency of a typical CCD given that all other properties are equal.

12. Draw a representative diagram of your own choosing that illustrates both types of CCD saturation on the same plot. Which of the types of CCD saturation actually is caused by charge overflow within a pixel?

13. What are the largest disadvantages of using a CCD operating in MPP mode?

14. OTCCDs are four-phase devices. Draw a voltage diagram similar to Figure 2.2 for such a device. How does an OTCCD pixel work? What does one do to readout an OTCCD as a normal CCD?

15. Make a drawing of the guide star region and the science region in an OTCCD. Detail how this combination is used to provide a tip-tilt correction to the science image.

16. Can you design a high-speed photometer using an L3CCD? (Chapters 4 and 5 will help you.)

17. Assuming that superconducting tunnel junctions were perfected and available on every telescope, discuss a good astronomical use of such a device. How would the observations be greatly improved over those available with a typical CCD?

18. Write down an equation, based on your knowledge of physics and optics, of how a coating applied to a CCD can act as an antireflection coating.

19. Read the paper by Opal (1988) and discuss how the choice of an A/D converter can affect your photometric precision and results obtained with a CCD.

20. Convince yourself that correlated double sampling is a differential measurement.

21. Why are CCDs really analog devices?

22. For accurate photometry, what gain value might you use? For imaging a scene with a large dynamic range what gain value might you use?

23. What are the most important features in a modern CCD controller?

3

Characterization of charge-coupled devices

Even casual users of CCDs have run across the terms read noise, signal-to-noise ratio, linearity, and many other possibly mysterious sounding bits of CCD jargon. This chapter will discuss the meanings of the terms used to characterize the properties of CCD detectors. Techniques and methods by which the reader can determine some of these properties on their own and why certain CCDs are better or worse for a particular application are discussed in the following chapters. Within the discussions, mention will be made of older types of CCDs. While these are generally not available or used anymore, there is a certain historical perspective to such a presentation and it will likely provide some amusement for the reader along the way.

One item to keep in mind throughout this chapter and in the rest of the book is that all electrons look alike. When a specific amount of charge is collected within a pixel during an integration, one can no longer know the exact source of each electron (e.g., was it due to a stellar photon or is it an electron generated by thermal motions within the CCD itself?). We have to be clever to separate the signal from the noise. There are two notable quotes to cogitate on while reading this text. The first is from an early review article on CCDs by Craig Mackay (1986), who states: "The only uniform CCD is a dead CCD." The second is from numerous discussions I have had with CCD users and it is: "To understand your signal, you must first understand your noise."

3.1 Quantum efficiency

The composition of a CCD is essentially pure silicon. This element is thus ultimately responsible for the response of the detector to various wavelengths of light. The wavelength dependence of silicon can be understood in an instant by glancing at Figure 3.1. Shown here is the length of silicon needed for a photon of a specific wavelength to be absorbed. Absorption length is defined as the distance for which 63% ($1/e$) of the incoming photons will be absorbed. Figure 3.1 clearly shows that, for light outside the range of about 3500 to over 8000 Å, the photons (1) pass right through the silicon, (2) get absorbed within the thin surface layers or gate structures, or (3) simply reflect off the CCD surface. At short wavelengths, 70% or more of the photons are reflected, and for very short wavelengths (as for long wavelengths) the CCD becomes completely transparent. Thus the quantum efficiency of a typical CCD device will approximately mirror the photon absorption curve for silicon. Shortward of ~ 2500 Å (for thinned devices) or about 25 Å (for thick devices) the detection probability for photons increases again. However, owing to their much higher energy, these photons lead to the production of

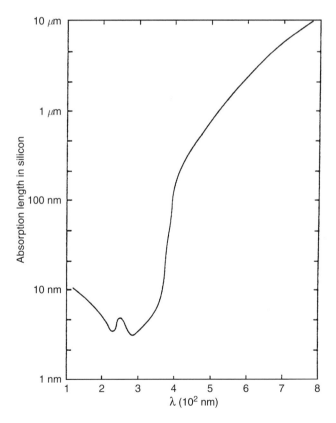

Fig. 3.1. The photon absorption length in silicon is shown as a function of wavelength in nanometers. From Reicke (1994).

multiple electron–hole pairs within the silicon and may also produce damage to the CCD itself (see Chapter 7).

CCD quantum efficiencies are therefore very dependent on the thickness of the silicon that intercepts the incoming photons. This relation between absorption probability and CCD thickness is why front-side illuminated (thick) devices are more red sensitive (the photons have a higher chance of absorption) and why they have lower overall (blue) QEs (since the gate structures can be close to or even exceed the necessary absorption depths of as small as only a few atomic layers). A few front-side CCDs have been produced with special gate structures that are transparent to incoming blue and UV photons. In thinned devices, the longest wavelength photons are likely to pass right through the CCD without being absorbed at all.

Figure 3.2 shows the quantum efficiencies for various imaging devices. Note that the y scale is logarithmic and the much superior QE provided by CCDs over previous detectors. Figure 3.3 shows a selection of modern CCD QEs. The large difference in QE that used to exist between thinned and thick CCDs is now mostly eliminated due to manufacturing processes

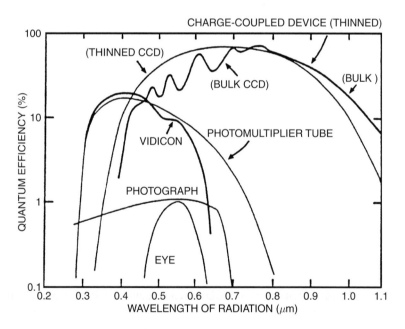

Fig. 3.2. QE curves for various devices, indicating why CCDs are a quantum leap above all previous imaging devices. The failure of CCDs at optical wavelengths shorter than about 3500 Å has been essentially eliminated via thinning or coating of the devices (see Figure 3.3).

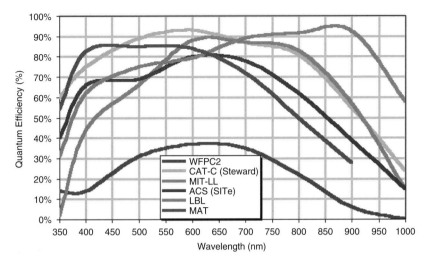

Fig. 3.3. QE curves for a variety of CCDs. WFPC2 is the second generation wide-field/planetary camera aboard HST, CAT-C is a new generation SITe CCD used in a mosaic imager at the University of Arizona's 90" telescope on Kitt Peak, MIT-LL is a CCD produced at the MIT Lincoln Laboratories and optimized for red observations, ACS is the Hubble Space Telescope Advanced Camera for Surveys SITe CCD, LBL is a Lawrence Berkeley Lab high resistivity, "deep depletion" CCD with high red QE, and MAT is a front-side, processed CCD showing high blue QE.

and coatings although other differences (such as location of peak QE, cosmic ray detection, etc.) remain. Quantum efficiency or QE curves allow one quickly to evaluate the relative collecting power of the device as a function of wavelength. Measured QE curves, such as in Figure 3.3 and those shown in the literature, are generally assumed to be representative of each and every pixel on the device, that is, all pixels of a given device are assumed to work identically and have the same wavelength response. This is almost true, but it is the "almost" that makes flat fielding of a CCD necessary. In addition, the QE curves shown or delivered with a particular device may only be representative of a "typical" device of the same kind, but they may not be 100% correct for the exact device of interest.

The quantum efficiency of a CCD is temperature sensitive especially in the red wavelength region. It has long been known that measurement of the QE at room temperature is a poor approximation to that which it will have when operated cold. Thus QE curves should be measured at or near the operating temperature at which the CCD will be used. As an example of the temperature sensitivity of the efficiency of a CCD, Figure 3.4 shows

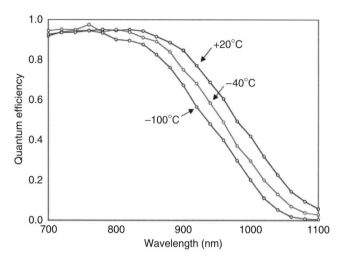

Fig. 3.4. Sensitivity of the quantum efficiency of a MIT/LL CCD for three operating temperatures. The blue sensitivity is little affected by a change in operating temperature but the red QE can change by a factor of two. The use of such devices requires a balance of higher operating temperature and keeping the dark current under control.

three QE measurements of the same CCD for temperatures of +20° C (~room temperature), −40°C, and −100°C (~operating temperature). Note that the variations are small at 8000 Å but increase to 20% at 9000–10 000 Å. The curves in this plot would lead one to the conclusion that operating a CCD at room temperature is the best thing to do. However, Section 3.5 will show us why this is not the case and a compromise between operating temperature (i.e., dark current) and delivered QE must be used.

A recent advance in the manufacture of CCDs is to use high resistivity silicon. Typical CCDs you have used have a resistivity of 20–200 ohm-cm or maybe up to 300 ohm-cm and are made on 10–40 micron epi.[1] The above resistivity and thickness values for Si wafers are fairly standard today and lend themselves to easy etching for thinning (10–20 micron final thickness). Starting with bulk silicon and a new process called the float-zone technique, resistivities up to 5000–10 000 ohm-cm are possible. Adding in a bias voltage to the optically transparent back-side substrate and using a thick layer of 45 to 350 micron epi, each pixel in a high resistivity CCD can be fully depleted resulting in very high red QE and deep pixel wells. Figure 3.5 shows the relationship between Si resistivity, depletion depth, and the bias voltage used.

[1] epi (pronounced "ep'-pea") is CCD lingo for epitaxial silicon, which is the Si wafer type used to make CCDs. An example was shown in Figure 2.8.

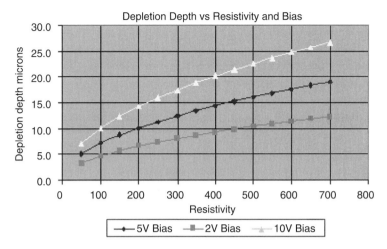

Fig. 3.5. Laboratory measurements of the depletion depth in a pixel vs. the CCD silicon resistivity for three different bias voltages. We see that one can deplete deeper (larger) pixels with higher resistivity silicon assuming the use of a larger bias voltage.

These "deep depletion" CCDs collect up to about 300 000 photoelectrons deep in a pixel where they are more likely to stay given the high resistivity. Higher resistivity silicon wafers require very special care in their production and much higher purity tolerances of the Si wafers, thus are more costly to produce. The internal Si lattice structures must be highly uniform and the level of unwanted impurities in the Si must be very near zero. Until the past few years, production of such Si was not possible and today the Lawrence Berkeley Lab (LBL) and MIT/Lincoln Labs are the leaders in making such devices. Figure 3.3 shows the superior red QE of a LBL high resistivity CCD. As we noted in our discussion of Figure 3.1, the thickness of a CCD is important in the QE it attains, and thus deep depletion devices have large well depths to aid in the improvement of their red QE. The resistivity of the Si and the deep pixels both come into play when one considers charge diffusion within a CCD. We will discuss charge diffusion in some detail below.

Placing an antireflection (AR) coating on a CCD (both for visible and near-UV light) increases the QE and extends the range of good QE well into the 3000 Å region. The need for AR coatings comes from the fact that silicon, like most metallic substances, is a good reflector of visible light. If you ever have a chance to hold a CCD, you will easily see just how well the surface does indeed reflect visible light. All the QE curves in Figure 3.3 have the overall shape expected based on the absorption properties of silicon as shown in Figure 3.1. Graphical illustrations of QE curves almost always include photon losses due to the gate structures, electron recombination within the bulk silicon itself, surface reflection, and, for very long or short wavelengths, losses due to the almost complete lack of absorption by the CCD. Given that all these losses are folded together into the QE value for each wavelength, it should be obvious that changes within the CCD structure itself (such as radiation damage or operating temperature changes) can cause noticeable changes in its quantum efficiency.

Measurement of the quantum efficiency of a CCD is usually performed with the aid of complicated laboratory equipment including well-calibrated photodiodes. Light at each wavelength is used to illuminate both the CCD and the photodiode, and the relative difference in the two readings is recorded. The final result of such an experiment is an absolute QE curve for the CCD (with respect to the calibrated diode) over the range of all measured wavelengths.

To measure a CCD QE curve yourself, a few possibilities exist. You may have access to a setup such as that described above. Measurements can also be made at the telescope itself. One good method of QE measurement for a CCD consists of employing a set of narrow-band filters and a few spectrophotometric standard stars. Performing such a task will provide a good relative QE curve and, if one knows the filter and telescope throughput well, a good absolute QE curve. A detailed reference as to what is involved in the measurement of a spectrophotometric standard star is provided by Tüg *et al.* (1977). When producing a QE curve by using the above idea, the narrow-band filters provide wavelength selection while the standard stars provide a calibrated light source. A less ambitious QE curve can be produced using typical broad-band (i.e., Johnson) filters, but the final result is not as good because of the large bandpasses and throughput overlap of some of the filters. In between a detailed laboratory setup and the somewhat sparse technique of using filters at the telescope, another method exists. Using an optics bench, a calibrated light source covering the wavelength range of interest, and some good laboratory skills, one can produce a very good QE curve for a CCD and can even turn the exercise into a challenging classroom project.

3.2 Charge diffusion

Once an electron is captured in a CCD pixel, the voltages applied during integration attempt to hold it in place. However, situations arise within a CCD pixel that provide a finite possibility for any given electron to wander out of its collection pixel and into a neighboring pixel. This process is called charge diffusion and until recently it was noted but of low significance compared with other noise and readout issues. Today CCDs are of excellent quality and have very low readout noise, good pixel registration on the array, and reside in high quality optical systems. These facts mean that CCD imaging now has the ability to show great detail of any optical aberrations and geometric distortions. Even items such as better mechanical tolerances in instrumentation can reveal noticeable focus variations as materials breathe with thermal changes. Given CCDs with deep pixel wells, large format front-side illuminated thinned devices, and the related improvements to modern astronomical instrumentation, the effects of charge diffusion on the point-spread function are noticeable.

A few ways in which charge diffusion can occur may be useful to discuss. Imagine a deep (geometrically long) pixel modeled after that which is shown in Figure 2.4 (also refer to Figure 3.5). Electrons produced by long wavelength photons are captured in the front-side illuminated pixel near the bottom, far from the applied voltages in the front gates. Thus the potential well for these electrons is more like a potential dip. Given the right circumstances, an electron can move into a neighboring pixel. Another example would be impurities in the silicon material the CCD was manufactured from. These variations in the Si lattice can slightly bend or slightly redirect the potential within a pixel and provide weak spots from which electron escape is possible. Ways to mitigate electron loss are the use of higher potential voltages (although this can lead to other issues such as logic glow or shorting on the array), higher resistivity Si (as discussed above) to more tightly hold the electrons (the Si lattice) in place, or to use small pixels (but these have lower red QE and small well depths). Again, we see that compromise and specific application come into play and can be tuned into the CCD as a part of its production process.

Charge diffusion often varies significantly across a CCD, especially thinned devices as they are not equal thickness everywhere. For example, the Hubble Space Telescope ACS wide field camera thinned CCD shows a variation in the core of the PSF, caused by charge diffusion, across the field of view. The variation is 30–40% at 5000 Å with larger variations at shorter wavelengths (Krist, 2004, HST ACS user manual). The effects of charge diffusion are not to be taken lightly. The ACS/WFC suffers a loss of about 0.5 magnitudes in its limiting magnitude at short wavelengths and near 0.2 magnitudes elsewhere. Charge diffusion is especially important in thinned devices that undersample the point-spread function.

3.3 Charge transfer efficiency

As we mentioned in the previous chapter, charge transfer efficiency or CTE is a measure of the fraction of the charge that is successfully transferred for each pixel transfer. CTE values of 0.999 995 or more are typical in good modern CCDs. For a CCD with 1024×1024 pixels, the charge collected in the last pixel readout has shifted 2048 times thus the CTE must be nearly 100% in order to preserve the charge in each pixel during readout. CTI (charge transfer inefficiency) is 1–CTE or numerically near 10^{-5} or 10^{-6} in value. CTI can be and usually is different in the vertical and horizontal directions. The loss in charge from a CCD pixel containing N electrons that is shifted 1024 times vertically and 1024 times horizontally is given by $L(e) = N(1024 * CTI(H) + 1024 * CTI(V))$ or, if a single CTI value is given, $L(e) = 2048 * N * CTI$. CCDs with poor CTE generally show charge tails in the direction opposite readout for bright stars. These tails are the charge left behind as the image is shifted out.

The standard method for measuring CTE is to use X-ray stimulation of a CCD with a Fe^{55} source. CCDs are good X-ray detectors (see Chapter 7) and for a specific X-ray source such as Fe^{55}, each X-ray photon collected produces ~ 1620 electrons.[1] A CCD is exposed to X-ray illumination and the resulting image readout. An X-ray transfer plot is made of signal in DN (y-axis) vs. running pixel number on the x-axis. Often hundreds of rows are summed together to increase the signal generated by the weak X-ray source. If the CCD has good CTE, a horizontal line will be seen at 1620 electrons (assuming a gain of 1.0). If the CTE is poor, this line starts at 1620 electrons (for rows close to the output amplifier) but tilts toward lower signal values for pixels further away from the output amplifier. This behavior indicates a loss of charge being transferred due to poor CTE. The CTE of a given CCD will generally degrade rapidly with decreasing operating temperature and is also a function of the clocking pulse shape and speed.

X-ray transfer techniques become imprecise for CTEs that are $> 0.999\,99$ and for CCDs with very large formats. For these CCDs, more sensitive CTE measurement techniques are required. A detailed discussion of the many intricacies of CTE in CCDs, how it is measured, and additional CTE measurement techniques are presented in Janesick (2001).

---

[1] Remember that for optical photon detection, one photon collected produces one photoelectron, regardless of its wavelength. For the much higher energy X-rays, a single photon collected produces multiple electrons in proportion to the photon's energy.

3.4 Readout noise

CCDs can be though of as having three noise regimes: read noise, shot noise, and fixed pattern noise. In astronomy, we speak of these as read noise limited, photon noise limited, and flat field uncertainties. A plot of the log of the standard deviation of the signal (*y*-axis) vs. the log of the signal itself (*x*-axis) for a CCD is called the photon transfer curve. Read noise (or any noise independent of signal level) sets a noise floor for a device. Upon illumination, photon or Poisson noise raises the sigma measured following a $N^{1/2}$ slope (see Chapter 4). Finally, for large signal values, pixel to pixel variations due to processing errors and photomask mis-alignment begin to dominate. This latter noise is proportional to the signal and rises with a slope of 1.0. Full well sets in at some very high illumination and the slope of the photon transfer curve turns over or breaks. We discuss read noise below and the total noise and flat fielding in the next chapter.

Readout noise, or just read noise, is usually quoted for a CCD in terms of the number of electrons introduced per pixel into your final signal upon readout of the device. Read noise consists of two inseparable components. First is the conversion from an analog signal to a digital number, which is not perfectly repeatable. Each on-chip amplifier and A/D circuit will produce a statistical distribution of possible answers centered on a mean value.[1] Thus, even for the hypothetical case of reading out the same pixel twice, each time with identical charge, a slightly different answer may be produced. Second, the electronics themselves will introduce spurious electrons into the entire process, yielding unwanted random fluctuations in the output. These two effects combine to produce an additive uncertainty in the final output value for each pixel. The average (one sigma) level of this uncertainty is the read noise and is limited by the electronic properties of the on-chip output amplifier and the output electronics (Djorgovski, 1984).[2]

The physical size of the on-chip amplifier, the integrated circuit construction, the temperature of the amplifier, and the sensitivity (generally near 1–4 µV/detected photon, i.e., collected photoelectron) all contribute to the read noise for a CCD. In this micro world, the values for electronic noise are highly related to the thermal properties of the amplifier, which in turn determines the sensitivity to each small output voltage. Amazing as it seems, the readout speed, and thus the rate at which currents flow through the on-chip

---

[1] The distribution of these values is not necessarily Gaussian (Merline & Howell, 1995).

[2] We note here that the level of the read noise measured, or in fact any noise source within a CCD, can never be smaller than the level of discretization produced by the A/D converter (see Sections 2.4 and 3.8).

amplifier, can cause thermal swings in the amplifier temperature, which can affect the resulting read noise level.[1] Generally, slower readout speeds produce lower read noise but this reduced readout speed must be weighed against the overall camera duty cycle. Small effects caused by amplifier heating can even occur between the readout of the beginning and end of a single CCD row, as the last charge packets pass through a slightly hotter circuit. Increasing the physical size of the already small microamplifiers can alleviate these small temperature swings, but larger amplifiers have a higher input capacitance, thereby lowering the sensitivity of the amplifier to small voltages.

Additional work on amplifier design, methods of clocking out pixels, and various semiconductor doping schemes can be used to increase the performance of the output electronics. Production techniques for integrated circuits have also contributed greatly to improved noise performance. An example of this improvement is the highly precise registration of each layer of a pixel's material and structure during production leading to more uniform electronic fields within the device and thus less resistive noise. CCD manufacturers invest large efforts into balancing these issues to produce very low read noise devices. Many details of the various aspects of read noise in CCDs are discussed by Janesick and Elliott (1992).

In the output CCD image, read noise is added into every pixel every time the array is readout. This means that a CCD with a read noise of 20 electrons will, on average, contain 20 extra electrons of charge in each pixel upon readout. High read noise CCDs are thus not very good to use if co-addition of two or more images is necessary. The final resultant image will not be quite as good as one long integration of the same total time, as each co-added image will add in one times the read noise to every pixel in the sum. However, for modern CCDs (see Section 4.4), read noise values are very low and are hardly ever the dominant noise with which one must be concerned. Good read noise values in today's CCDs are in the range of 10 electrons per pixel per read or less. These values are far below read noise levels of ten years ago, which were as high as 50–100 electrons, and even those are well down from values of 300–500 or more electrons/pixel/read present in the first astronomical CCDs.

In Section 4.3, we will discuss a simple method by which one may determine for oneself the read noise of a given CCD. This determination can be performed with any working CCD system and does not require special equipment, removal of the CCD from the camera dewar, or even removal from the telescope.

---

[1] Figure 6.8a, spectrum a, shows this effect for the first few columns in each row.

3.5 Dark current

Every material at a temperature much above absolute zero will be subject to thermal noise within. For silicon in a CCD, this means that when the thermal agitation is high enough, electrons will be freed from the valence band and become collected within the potential well of a pixel. When the device is readout, these dark current electrons become part of the signal, indistinguishable from astronomical photons. Thermal generation of electrons in silicon is a strong function of the temperature of the CCD, which is why astronomical use generally demands some form of cooling (McLean, 1997b). Figure 3.6 shows a typical CCD dark current curve, which relates the amount of thermal dark current to the CCD operating temperature. Within the figure the theoretical relation for the rate of thermal electron production is given.

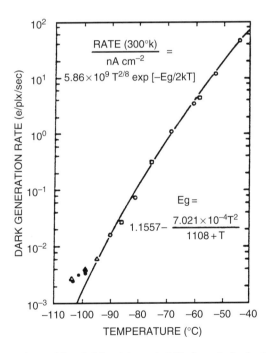

Fig. 3.6. Experimental (symbols) and theoretical (line) results for the dark current generated in a typical three-phase CCD. The rate of dark current, in electrons generated within each pixel every second, is shown as a function of the CCD operating temperature. E_g is the band gap energy for silicon. From Robinson (1988a).

Dark current for a CCD is usually specified as the number of thermal electrons generated per second per pixel or as the actual current generated per area of the device (i.e., picoamps cm^{-2}). At room temperature, the dark current of a typical CCD is near 2.5×10^4 electrons/pixel/second. Typical values for properly cooled devices range from 2 electrons per second per pixel down to very low levels of approximately 0.04 electrons per second for each pixel. Although 2 electrons of thermal noise generated within a pixel every second sounds very low, a typical 15 minute exposure of a faint astronomical source would include 1800 additional (thermal) electrons within each CCD pixel upon readout. These additional charges cannot, of course, be uniquely separated from the photons of interest after readout. The dark current produced in a CCD provides an inherent limitation on the noise floor of a CCD. Because dark noise has a Poisson distribution, the noise actually introduced by thermal electrons into the signal is proportional to the square root of the dark current (see Section 4.4).

Cooling of CCDs is generally accomplished by one of two methods. The first, and usually the one used for scientific CCDs at major observatories, is via the use of liquid nitrogen (or in some cases liquid air). The CCD and associated electronics (the ones on or very near the actual CCD itself, called the head electronics) are encased in a metal dewar under vacuum. Figure 3.7 shows a typical astronomical CCD dewar (Brar, 1984; Florentin-Nielsen, Anderson, & Nielsen, 1995). The liquid nitrogen (LN2) is placed in the dewar and, although not in direct physical contact with the CCD, cools the device to temperatures of near $-100°$ C. Since LN2 itself is much colder than this, CCDs are generally kept at a constant temperature ($\pm0.1°$ C) with an on-board heater. In fact, the consistency of the CCD temperature is very important as the dark current is a strong function of temperature (Figure 3.6) and will vary considerably owing to even modest changes in the CCD temperature.

A less expensive and much less complicated cooling technique makes use of thermoelectric cooling methods. These methods are employed in essentially all "off-the-shelf" CCD systems and allow operation at temperatures of -20 to $-50°$ C or so, simply by plugging the cooler into an electrical outlet. Peltier coolers are the best known form of thermoelectric cooling devices and are discussed in Martinez & Klotz (1998). CCD operation and scientific quality imaging at temperatures near $-30°$ C is possible, even at low light levels, due to advances in CCD design and manufacturing techniques and the use of multipinned phase operation (see Chapter 2). Other methods of cooling CCDs that do not involve LN2 are discussed in McLean (1997a).

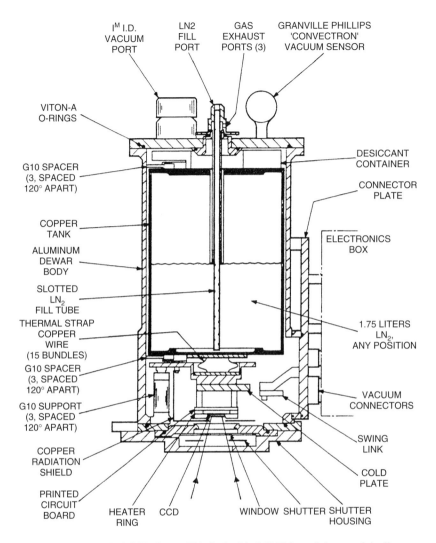

Fig. 3.7. A typical CCD dewar. This is the Mark-II Universal dewar originally produced in 1984 at Kitt Peak National Observatory. The dewar held 1.75 liters of liquid nitrogen providing a CCD operating time of approximately 12 hours between fillings. This dewar could be used in up-looking, down-looking, and side-looking orientations. From Brar (1984).

The amount of dark current a CCD produces depends primarily on its operating temperature, but there is a secondary dependence upon the bulk properties of the silicon used in the manufacture. Even CCDs produced on the same silicon wafer can have slightly different dark current properties.

Today's CCDs are made from high purity epi wafers produced with low occurrences of integrated circuit error. These factors have greatly reduced many of the sources of dark current even at warmer temperatures. As with most of the noise properties of a given CCD, custom tailoring the CCD electronics (such as the bias level and the readout rate) can produce much better or much worse overall dark current and noise performance.

3.6 CCD pixel size, pixel binning, full well capacity, and windowing

This section is a combination of a few related topics concerning the amount of charge that can be stored within a given pixel during an integration. We have seen that CCD thinning, MPP operation, and small physical pixel size all place limitations on the total number of electrons that can be collected within a pixel. The general rule of thumb is that the physically larger the pixel (both in area and in thickness) the more charge that it can collect and store.

The amount of charge a pixel can hold in routine operation is termed its full well capacity. A Kodak CCD with 9-micron pixels (meaning 9 microns on a side for the projected area, but giving no indication of the thickness of the CCD) operating in MPP mode has a full well capacity per pixel of 85 000 electrons. In contrast, a SITe CCD with 24-micron pixels can have a full well capacity per pixel of over 350 000 electrons. CCDs have been produced today which have 1 million electron well depths per pixel. While this value is highly desirable, it is not without compromise. Keep in mind our above discussion of charge diffusion.

When we discussed the method by which a CCD is readout (Chapter 2) it was stated that each row is shifted in turn into the output register and then digitized, and the resulting DN value is sent off to the computer. During this process, each pixel's value is increased on average by one times the read noise. If we could add up the charge within say 4 pixels before they are digitized, we would get a final signal level equal to ~ 4 times each single pixel's value, but only one times the read noise. This process is called on-chip binning and, if selected, occurs prior to readout within the CCD output register (Smith, 1990b; Merline & Howell, 1995). Pixels can be binned (summed) in both vertical and horizontal directions. "On-chip" means that the accumulated charge from each pixel involved in the binning is brought together and summed before the process of A/D conversion occurs. This summing process is done in the output register and is limited by the size of the "pixels" within this register. Generally, the output register pixels can hold five to ten times the charge of a single active pixel. This deeper full well capacity of the output register pixels allows pixel summing to take place.

Older CCD systems that allowed on-chip binning had plug boards mounted on the sides of the dewar. Certain combinations of the plug wires produced different on-chip binning patterns and users could change these to suit their needs. Today, most CCDs have the ability to perform pixel summing as a software option (Leach, 1995). Binning terminology states that normal operation (or "high resolution" as it is called by many low-cost CCDs) is a readout of the CCD in which each pixel is read, digitized, and stored. This is called 1×1 binning or unbinned operation. Binning of 2×2 would mean that an area of four adjacent pixels will be binned or summed on-chip within the output register during readout, but before A/D conversion. The result of this binning operation will produce only one "superpixel" value, which is digitized and stored in the final image; the original values in each of the four summed pixels are lost forever. Mainly for spectroscopic operation, binning of 3×1 is commonly used, with the 3 being in the direction perpendicular to the dispersion. Binning of CCD pixels decreases the image resolution, usually increases the final signal-to-noise value of a measurement, and reduces the total readout time and final image size. For example, a 1024×1024 CCD binned 2×2 will have a final image size of only 512×512 pixels and the readout time will be reduced by about a factor of four.

Pixel binning gives flexibility to the user for such applications as (using a high binning factor) quick readout for focus tests, nights with poor seeing, or very low surface brightness observations. Spectroscopic observations with a CCD, high spatial resolution imaging, or bright object observations will benefit from the use of a low binning factor. Binning factors that are very large (say 40×40 pixels) might be of use in some rare cases, but they will be limited by the total amount of charge one can accumulate in a single superpixel of the output register.

A related function available with some CCDs is "windowing." Windowing allows the user to choose a specific rectangular region (or many regions) within the active area of the CCD to be readout upon completion of the integration. The CCD window is often specified by providing the operating software with a starting row and column number and the total number of x, y pixels to use. For example, using a 2048×2048 CCD to make high-speed imaging observations would be difficult, but windowing the CCD to use only the first 512 rows and columns $(0, 0, 512, 512)$ allows for much faster readout and requires far less storage for the image data. The use of subregion readout for astronomical CCDs is often the heart of fast imaging cameras such as UltraCam and OPTIC. New generation OTCCDs allow for fast readout via the use of not only fast readout electronics (available to all modern CCDs) but by having no single CCD larger than about 512×512 pixels.

Of course, the object of interest must be positioned within these first 512 rows and columns, and not at the center of the CCD as may be usual. Other applications of CCD windowing would include choosing a cosmetically good subregion of a large CCD or only a rectangular strip to readout from a larger square CCD, when making spectroscopic observations. CCD windowing is independent of any on-chip binning, and one can both window and bin a CCD for even more specific observational needs.

3.7 Overscan and bias

In an attempt to provide an estimate of the value produced by an empty or unexposed pixel within a CCD, calibration measurements of the bias level can be used.[1] Bias or zero images allow one to measure the zero noise level of a CCD. For an unexposed pixel, the value for zero collected photoelectrons will translate, upon readout and A/D conversion, into a mean value with a small distribution about zero.[2] To avoid negative numbers in the output image,[3] CCD electronics are set up to provide a positive offset value for each accumulated image. This offset value, the mean "zero" level, is called the bias level. A typical bias level might be a value of 400 ADU (per pixel), which, for a gain of $10\,e^-/\text{ADU}$, equals 4000 electrons. This value might seem like a large amount to use, but historically temporal drifts in CCD electronics due to age, temperature, or poor stability in the electronics, as well as much higher read noise values, necessitated such levels.

---

[1] For more on bias frames and their use in the process of CCD image calibration, see Chapter 4.

[2] Before bias frames, and in fact before any CCD frame is taken, a CCD should undergo a process known as "wiping the array." This process makes a fast read of the detector, without A/D conversion or data storage, in order to remove any residual dark current or photoelectron collection that may have occurred during idle times between obtaining frames of interest.

[3] Representation of negative numbers requires a sign bit to be used. This bit, number 15 in a 16-bit number, is 0 or 1 depending on whether the numeric value is positive or negative. For CCD data, sacrificing this bit for the sign of the number leaves one less bit for data, thus reducing the overall dynamic range. Therefore, most good CCD systems do not make use of a sign bit. One can see the effects of having a sign bit by viewing CCD image data of high numeric value but displayed as a signed integer image. For example, a bright star will be represented as various grey levels, but at the very center (i.e., the brightest pixels) the pixel values may exceed a number that can be represented by 14 bits (plus a sign). Once bit 15 is needed, the signed integer representation will be taken by the display as a negative value and the offending pixels will be displayed as black. This is due to the fact that the very brightest pixel values have made use of the highest bit (the sign bit) and the computer now believes the number is negative and assigns it a black (negative) greyscale value. This type of condition is discussed further in Appendix.

To evaluate the bias or zero noise level and its associated uncertainty, specific calibration processes are used. The two most common ones are: (1) overscan regions produced with every object frame and (2) usage of bias frames. Bias frames amount to taking observations without exposure to light (shutter closed), for a total integration time of 0.000 seconds. This type of image is simply a readout of the unexposed CCD pixels through the on-chip electronics, through the A/D converter, and then out to the computer producing a two-dimensional bias or zero image.

Overscan strips, as they are called, are a number of rows or columns (usually 32) or both that are added to and stored with each image frame. These overscan regions are not physical rows or columns on the CCD device itself but additional pseudo-pixels generated by sending additional clock cycles to the CCD output electronics. Both bias frames and overscan regions are techniques that allow one to measure the bias offset level and, more importantly, the uncertainty of this level.

Use of overscan regions to provide a calibration of the zero level generally consists of determining the mean value within the overscan pixels and then subtracting this single number from each pixel within the CCD object image. This process removes the bias level pedestal or zero level from the object image and produces a bias-corrected image. Bias frames provide more information than overscan regions, as they represent any two-dimensional structure that may exist in the CCD bias level. Two-dimensional (2-D) patterns are not uncommon for the bias structure of a CCD, but these are usually of low level and stable with time. Upon examination of a bias frame, the user may decide that the 2-D structure is nonexistent or of very low importance and may therefore elect to perform a simple subtraction of the mean bias level value from every object frame pixel. Another possibility is to remove the complete 2-D bias pattern from the object frame using a pixel-by-pixel subtraction (i.e., subtract the bias image from each object image). When using bias frames for calibration, it is usually best to work with an average or median frame composed of many (10 or more) individual bias images (Gilliland, 1992). This averaging eliminates cosmic rays,[1] read noise variations, and random fluctuations, which will be a part of any single bias frame.

Variations in the mean zero level of a CCD are known to occur over time and are usually slow drifts over many months or longer, not noticeable changes from night to night or image to image. These latter types of changes

---

[1] Cosmic rays are not always cosmic! They can be caused by weakly radioactive materials used in the construction of CCD dewars (Florentin-Nielsen, Anderson, & Nielsen, 1995).

indicate severe problems with the readout electronics and require correction before the CCD image data can be properly used.

Producing a histogram of a typical averaged bias frame will reveal a Gaussian distribution with the mean level of this distribution being the bias level offset for the CCD. We show an example of such a bias frame histogram in Figure 3.8. The width of the distribution shown in Figure 3.8 is related to the read noise of the CCD (caused by shot noise variations in the CCD electronics (Mortara & Fowler, 1981)) and the device gain by the following expression:

$$\sigma_{\mathrm{ADU}} = \frac{\text{Read noise}}{\text{Gain}}.$$

Note that σ is used here to represent the width (FWHM) of the distribution not the usual definition for a Gaussian shape. For example, in Figure 3.8, $\sigma = 2$ ADU.

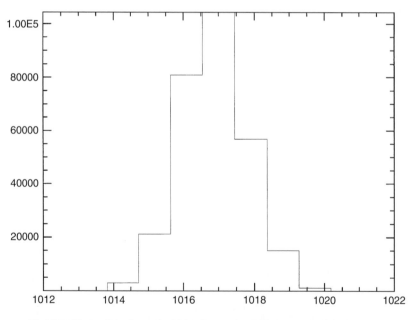

Fig. 3.8. Histogram of a typical bias frame showing the number of pixels vs. each pixel ADU value. The mean bias level offset or pedestal level in this Loral CCD is near 1017 ADU, and the distribution is very Gaussian in nature with a FWHM value of near 2 ADU. This CCD has a read noise of 10 electrons and a gain of $4.7\,e^-/\mathrm{ADU}$.

3.8 CCD gain and dynamic range

The gain of a CCD is set by the output electronics and determines how the amount of charge collected in each pixel will be assigned to a digital number in the output image. Gain values are usually given in terms of the number of electrons needed to produce one ADU step within the A/D converter. Listed as electrons/Analog-to-Digital Unit (e⁻/ADU), common gain values range from 1 (photon counting) to 150 or more. One of the major advantages of a CCD is that it is linear in its response over a large range of data values. Linearity means that there is a simple linear relation between the input value (charge collected within each pixel) and the output value (digital number stored in the output image).

The largest output number that a CCD can produce is set by the number of bits in the A/D converter. For example, if you have a 14-bit A/D, numbers in the range from zero to 16 383 can be represented.[1] A 16-bit A/D would be able to handle numbers as large as 65 535 ADU.

Figure 3.9 provides a typical example of a linearity curve for a CCD. In this example, we have assumed a 15-bit A/D converter capable of producing output DN values in the range of 0 to 32 767 ADU, a device gain of 4.5 e⁻/ADU, and a pixel full well capacity of 150 000 electrons. The linearity curve shown in Figure 3.9 is typical for a CCD, revealing that over most of the range the CCD is indeed linear in its response to incoming photons. Note that the CCD response has the typical small bias offset (i.e., the output value being nonzero even when zero incident photons occur), and the CCD becomes nonlinear at high input values. For this particular CCD, nonlinearity sets in near an input level of 1.17×10^5 photons (26 000 ADU), a number still well within the range of possible output values from the A/D.

As we have mentioned a few times already in this book, modern CCDs and their associated electronics provide high-quality, low-noise output. Early CCD systems had read noise values of 100 times or more of those today and even five years ago, a read noise of 15 electrons was respectable. For these systems, deviations from linearity that were smaller than the read noise were rarely noticed, measurable, or of concern. However, improvements that have lowered the CCD read noise provide an open door to allow other subtleties to creep in. One of these is device nonlinearities. Two types of nonlinearity are quantified and listed for today's A/D converters. These are integral nonlinearity and differential nonlinearity. Figure 3.10 illustrates these two types of A/D nonlinearity.

---

[1] The total range of values that a specific number of bits can represent equals $2^{(\text{number of bits})}$, e.g., $2^{14} = 16\,384$. CCD output values are zero based, that is, they range from 0 to $2^{(\text{number of bits})} - 1$.

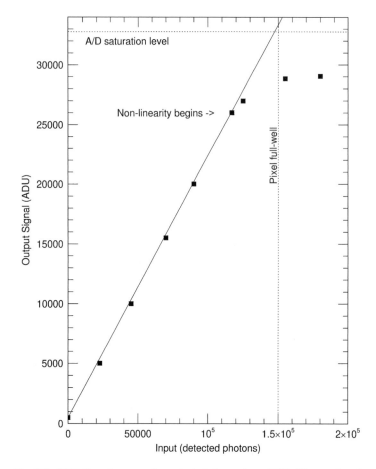

Fig. 3.9. CCD linearity curve for a typical three-phase CCD. We see that the
device is linear over the output range from 500 ADU (the offset bias level of
the CCD) to 26 000 ADU. The pixel full well capacity is 150 000 electrons and the
A/D converter saturation is at 32 767 ADU. In this example, the CCD nonlinearity
is the limiting factor of the largest usable output ADU value. The slope of the
linearity curve is equal to the gain of the device.

A/D converters provide stepwise or discrete conversation from the input
analog signal to the output digital number. The linearity curve for a CCD is
determined at various locations and then drawn as a smooth line approximation
of this discrete process. Differential nonlinearity (DNL) is the maximum
deviation between the line approximation of the discrete process and the A/D
step used in the conversion. DNL is often listed as ±0.5 ADU meaning
that for a given step from say 20 to 21 ADU, fractional counts of 20.1,

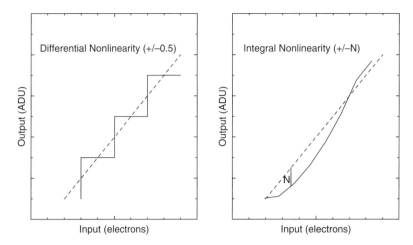

Fig. 3.10. The two types of CCD nonlinearity are shown here in cartoon form. Differential nonlinearity (left) comes about due to the finite steps in the A/D conversion process. Here we see that the linearity curve (dashed line) cuts through each step at the halfway point yielding a DNL of ±0.5 ADU. Integral nonlinearity (right) is more complex and the true linearity curve (solid line) may have a simple or complex shape compared with the measured curve (dashed line). A maximum deviation (N) is given as the INL value for an A/D and may occur anywhere along the curve and be of either sign. Both plots have exaggerated the deviation from linearity for illustration purposes.

20.2, etc. up to 20.499 99 will yield an output value of 20 while those of 20.5, 20.6, etc. will yield an output value of 21. Astronomers call this type of nonlinearity digitization noise and we discuss it in more detail below. Integral nonlinearity (INL) is of more concern as it is the maximum departure an A/D will produce (at a given convert speed) from the expected linear relationship. A poor quality A/D might have an INL value of 16 LSB (least significant bits). The value of 16 LSB means that this particular A/D has a maximum departure from linearity of 4 bits ($2^4 = 16$) throughout its full dynamic range. If the A/D is a 16-bit device and all 16 bits are used, bits 0–3 will contain any INL at each ADU step. If one uses the top 12 bits, then bits 4–7 are affected.

How the INL comes into play for an observer is as follows. For a gain of say 5 electrons/ADU, an INL value of 16 can cause a nonlinear deviation of up to 80 electrons in the conversion process at its maximum deviation step (see Figure 3.10). Thus, at the specific A/D step that has the maximum deviation, an uncertainty of ±80 electrons will occur but be unknown to the user. This is a very unacceptable result for astronomy, but fine for digital cameras or photocopiers that usually have even higher values of INL caused by their very fast readout (conversion) speeds.

A good A/D will have an INL value near 2–2.5 LSB or, for the above example, a maximum deviation of only 10 electrons. While this sounds bad, a 16-bit A/D can represent 65 535 values making the 10 electrons only a 0.02% nonlinearity. However, a 12-bit A/D, under similar circumstances, would have a 0.2% nonlinearity. The lesson here is to use a large dynamic range (as many bits as possible) to keep the nonlinearity as small as possible. We can now obtain A/D converters with low values for INL and which have 18 bits of resolution. So for a given modern CCD, nonlinearity is usually a small but nonzero effect.

Three factors can limit the largest usable output pixel value in a CCD image: the two types of saturation that can occur (A/D saturation and exceeding a pixel's full well capacity; see Sections 2.2.4 and 2.4) and nonlinearity. For the CCD in the example shown in Figure 3.9, A/D saturation would occur at an output value of $32\,767 \cdot 4.5 = 147\,451$ input photons. The pixel full well capacity is 150 000 electrons; thus pixel saturation will occur at a value of 33 333 ADU (150 000/4.5). Both full well and A/D saturations would produce noticeable effects in the output data such as bleeding or flat-topped stars. This particular example, however, illustrates the most dangerous type of situation that can occur in a CCD image. The nonlinear region, which starts at 26 000 ADU, is entered into before either type of saturation can occur. Thus, the user could have a number of nonlinear pixels (for example the peaks of bright stars) and be completely unaware of it. No warning bells will go off and no flags will be set in the output image to alert the user to this problem. The output image will be happy to contain (and the display will be happy to show) these nonlinear pixel values and the user, if unaware, may try to use such values in the scientific analysis.

Thus it is very important to know the linear range of your CCD and to be aware of the fact that some pixel values, even though they are not saturated, may indeed be within the nonlinear range and therefore unusable. Luckily, most professional grade CCDs reach one of the two types of saturation before they enter their nonlinear regime. Be aware, however, that this is almost never the case with low quality, inexpensive CCD systems that tend to use A/Ds with fewer bits, poor quality electronics, or low grade (impure) silicon. Most observatories have linearity curves available for each of their CCDs and some manufacturers include them with your purchase.[1] If uncertain of the linear range of a CCD, it is best to measure it yourself.

---

[1] A caution here is that the supplied linearity curve may only be representative of your CCD.

One method of obtaining a linearity curve for a CCD is to observe a field of stars covering a range of brightness. Obtain exposures of say 1, 2, 4, 8, 16, etc. seconds, starting with the shortest exposure needed to provide good signal-to-noise ratios (see Section 4.4) for most of the stars and ending when one or more of the stars begins to saturate. Since you have obtained a sequence that doubles the exposure time for each frame, you should also double the number of incident photons collected per star in each observation. Plots of the output ADU values for each star versus the exposure time will provide you with a linearity curve for your CCD.

A common, although not always good, method of determining the value to use for the CCD gain, is to relate the full well capacity of the pixels within the array to the largest number that can be represented by your CCD A/D converter. As an example, we will use typical values for a Loral 512×1024 CCD in current operation at the Royal Greenwich Observatory. This CCD has 15-micron pixels and is operated as a back-side illuminated device with a full well capacity of 90 000 electrons per pixel. Using a 16-bit A/D converter (output values from 0 to 65 535) we could choose the gain as follows. Take the total number of electrons a pixel can hold and divide it by the total ADU values that can be represented: $90\,000/65\,536 = 1.37$. Therefore, a gain choice of $1.4\,e^-$/ADU would allow the entire dynamic range of the detector to be represented by the entire range of output ADU values. This example results in a very reasonable gain setting, thereby allowing the CCD to produce images that will provide good quality output results.

As an example of where this type of strategy would need to be carefully thought out, consider a CCD system designed for a spacecraft mission in which the A/D converter only had 8 bits. A TI CCD was to be used, which had a full well capacity of 100 000 electrons per pixel. To allow imagery to make use of the total dynamic range available to the CCD, a gain value of 350 ($\sim 100\,000/2^8$) e^-/ADU was used. This gain value certainly made use of the entire dynamic range of the CCD, allowing images of scenes with both shadow and bright light to be recorded without saturation. However, as we noted before, each gain step is discrete, thereby making each output ADU value uncertain by $\pm$ the number of electrons within each A/D step. A gain of $350\,e^-$/ADU means that each output pixel value has an associated uncertainty of upto ~ 1 ADU, which is equal to, in this case, upto 350 electrons, a large error if precise measurements of the incident flux are desired. The uncertainty in the final output value of a pixel, which is caused by the discrete steps in the A/D output, is called digitization noise and is discussed in Merline & Howell (1995).

To understand digitization noise let us take, as an example, a CCD that can be operated at two different gain settings. If we imagine the two gain values to be either 5 or 200 e$^-$/ADU and that a particular pixel collects 26 703 electrons (photons) from a source, we will obtain output values of 5340 and 133 ADU respectively. Remember, A/D converters output only integer values and so any remainder is lost. In this example, 3 and 103 electrons respectively are lost as the result of the digitization noise of the A/D. More worrisome than this small loss of incident light is the fact that while each ADU step in the gain equals 5 e$^-$/ADU case can only be incorrect by < 5 electrons, the gain equals 200 e$^-$/ADU case will be uncertain by upto 200 electrons in each output ADU value. Two hundred electrons/pixel may not seem like much but think about trying to obtain a precise flux measurement for a galaxy that covers thousands of pixels on a CCD image or even a star that may cover tens of pixels. With an error of 200 electrons/pixel multiplied by tens or many more pixels, the value of a galaxy's surface brightness at some location or similarly a stellar magnitude would be highly uncertain.

The gain of a particular CCD system is set by the electronics and is generally not changeable by the user or there may be but a few choices available as software options. How A/D converters actually determine the assignment of the number to output for each pixel and whether the error in this choice is equally distributed within each ADU step is a detailed matter of interest but lies outside the scope of this book. A discussion of how the digitization noise affects the final output results from a CCD measurement is given in Merline & Howell (1995) and a detailed study of ADCs used for CCDs is given in Opal (1988).

Major observatories provide detailed information to a potential user (generally via internal reports or web pages) as to which CCDs are available. Table 3.1 gives an example of some of the CCDs in use in various instruments at the European Southern Observatory (ESO) in Chile. When planning an observational program, one must not only decide on the telescope and instrument to use, but must also be aware of the CCD(s) available with that instrument. The properties of the detector can be the most important factor in determining the success or failure of an observational project. Thus, some care must be taken in deciding which CCD, with its associated properties, you should use to accomplish your science objectives.

In our above discussion of the gain of a CCD, we mentioned the term dynamic range a few times but did not offer a definition. The dynamic range of any device is the total range over which it operates or for which it is sensitive. For audio speakers this number is usually quoted in decibels, and

Table 3.1. *Some CCDs available at the European Southern Observatory (ESO)*

Instrument	Telescope	Type	CCD	Size (pixels)	Pixel Size (microns)	Pixel Scale (arcsec)	Readout Time (seconds)	Gain (e⁻/ADU)	Read Noise electrons	Notes
EMMI	3.6-m NTT	Spectrograph Red channel	MIT/LL	2048 × 4096	15	0.17	18–48	1.4	4	Thinned, back-side
EMMI	3.6-m NTT	Spectrograph Blue channel	MIT/LL	1024 × 1024	24	0.37	40	1.4, 2.8	7	Thinned, back-side
FORS2	8-m VLT	Imager/ Spectrograph	MIT/LL	2048 × 4096	15	0.12	40	1.1	6	Deep Depletion, Red optimized
OmegaCam	2.6-m VST	Wide-Field imager	E2V	2048 × 4096	15	0.21	45	3	5	

this tradition has been used for CCDs as well. Keeping to the idea of decibels as a measure of the dynamic range of a CCD, we have the expression

$$D(dB) = 20 \times \log_{10}(\text{full well capacity/read noise}).$$

Thus a CCD with a full well capacity of 100 000 electrons per pixel and a read noise of 10 electrons would have a dynamic range of 80 dB. A more modern (and more useful) definition for the dynamic range of a CCD is simply the ratio of the (average) full well capacity of a pixel to the read noise of the device, namely

$$D = (\text{full well capacity/read noise}).$$

In the example above, $D = 10\,000$.

3.9 Summary

This chapter has concentrated on defining the terminology used when discussing CCDs. The brief nature of this book does not allow the many more subtle effects, such as deferred charge, cosmic rays, or pixel traps, to be discussed further nor does it permit any discussion of the finer points of each of the above items. The reader seeking a deeper understanding of the details of CCD terminology (a.k.a., someone with a lot of time on his or her hands) is referred to the references given in this chapter and the detailed reading list in Appendix A. Above all, the reader is encouraged to find some CCD images and a workstation capable of image processing and image manipulation and to spend a few hours of time exploring the details of CCDs for themselves.

As a closing thought for this chapter, Table 3.2 provides a sample of the main properties of two early astronomical CCDs and a few modern devices. The sample shown tries to present the reader with an indication of the typical properties exhibited by CCDs. Included are those of different dimension, of different pixel size, having front and back illumination, cooled by LN2 or thermoelectrically, and those available from different manufacturers. Information such as that shown in Table 3.2 can be found at observatory websites and in greater detail at CCD manufacturers' websites. Most have readily available data sheets for the entire line of CCDs they produce. Each example for a given CCD in Table 3.2 is presented to show the range of possible properties and does not imply that all CCDs made by a given company are of the listed properties. Most manufacturers produce a wide variety of device types. Appendix B provides a listing of useful CCD websites.

Table 3.2. *Typical Properties of Two Old and Six Modern Example CCDs*

	RCA	TI	Kodak	E2V	SITe	Sarnoff	STA (WIYN)	MIT/LL
Pixel Format	320 × 512	800 × 800	2048 × 2048	2048 × 4608	2048 × 2048	600 × 2400	3840 × 3952 OT	2048 × 4096
Pixel Size (microns)	30	15	9	13	12	13	12	15
Detector Size (mm)	10 × 15	12 × 12	18 × 18	27 × 62	25 × 25	6 × 25	50	31 × 62
Pixel Full Well (e⁻)	350 000	50 000	100000	150000	110000	> 20000	> 70000	> 200000
Illumination	Front	Back	Front	Back	Back	Back	Back	Back
Peak QE (%) / Wavelength (A)	70/4500	70/6500	45/6500	90/5000	85/6500	99/6600	96/5500	95/7700
Read Noise (e⁻)	80	15	15	3	6	6	< 5	2.5
CTE	0.99995	0.999985	0.99998	0.999995	0.99999	0.99999	0.999998	0.999995
Operating Temp (C)	−100	−120	−30	−85	−85	−60	−60	−110
Typical Gain used (e⁻/ADU)	13.5	5	5	1.5	3	5	1.5	1.37

3.10 Exercises

1. Using only the data presented in Figures 3.1 and 3.2, draw a quantum efficiency curve expected for a typical CCD. Why might real QE curves be different?

2. Discuss two major reasons why CCDs are better detectors than the human eye. Are there instances in which the eye is a better detector? What "type" of A/D converter does the eye have?

3. Design a detailed observing plan or laboratory experiment that would allow you to measure the quantum efficiency of a CCD. Discuss the specific light sources (astronomical or laboratory) you might use and over what band-passes you can work. How accurate a result would you expect?

4. Why is charge diffusion important to consider in a deep depletion CCD? Using the standard physics equation for diffusion, can you estimate the area over which electrons from one pixel will spread in a CCD as a function of time? (You will have to look up the properties of bulk silicon and keep in mind the operating temperatures and voltages.)

5. Make a list of the various CCD properties that contribute to CTI. For each, discuss a method for mitigation.

6. When does read noise get introduced into each pixel of a CCD during the readout process? How could you design a CCD to have zero read noise?

7. A CCD has a typical background level of 100 ADUs per pixel and a read noise of 6 electrons rms. An image is obtained that contains only read noise. What range of values would one expect to find in any pixel on the array? How would these values be distributed around the 100 ADU value?

8. Using the data presented in Figure 3.6, estimate the dark current for that CCD at room temperature. Given your answer, how do video or digital cameras record scenes that are not saturated by thermal noise?

9. Estimate the dark current for the CCD illustrated in Figure 3.6 at liquid nitrogen temperatures, at $-120\,°C$, at dry ice temperatures, and if using a thermoelectric cooler. What level of dark current is acceptable?

10. Do the numbers discussed in Section 3.6 concerning pixel size and full well capacity agree with your calculations from Question 8 in Chapter 2?

11. Discuss an observational application that might require CCD windowing and one that might require CCD binning. What limits the practical use of CCD binning on any given chip?

12. Detail the difference between overscan and bias. How are each related to a "zero" or bias image?

13. Why do CCDs have a bias level at all?

14. What is so important about a device being linear in its response to light?

15. For a CCD with a full well capacity of 90 000 electrons per pixel and a 12-bit A/D, what gain value would you choose and why? How might your choice change if the CCD became nonlinear at 65 000 electrons?

16. Design a detailed observing plan or laboratory experiment that would allow you to measure the linearity of a CCD. Discuss the specific light sources (astronomical or laboratory) you might use and the sequence of integrations you would take. What measurements would you make from the collected images? Over what band-passes would you work and how accurate a result would you expect?

17. Which type of nonlinearity is more acceptable in a CCD for spectroscopic observations? For photometric observations? What would the output from an A/D converter look like if the DNL was 0.1 instead of 0.5? What if the INL was 32?

18. What is "digitization noise" and under what conditions is it undesirable?

19. Using Table 3.2, discuss the best CCD to use for spectroscopic observations of sources with faint continua but very bright emission lines. What is the best CCD to use if you were attempting to measure very weak stellar absorption lines?

20. Compare the dynamic range of a CCD to that of a typical sub-woofer speaker. Compare it to a police-car siren.

4

CCD imaging

This chapter will deal with the most basic use of a CCD, that of direct imaging. We will discuss a few more preliminaries such as flat fields, the calculation of gain and read noise for a CCD, and how the signal-to-noise value for a measurement is determined. The chapter then continues by providing a brief primer on the use of calibration frames in standard two-dimensional CCD data image reduction. Finally, we cover some aspects of CCD imaging itself including applications of wide-field imaging with CCD mosaics and CCD drift scanning.

4.1 Image or plate scale

One of the basic parameters of importance to a CCD user is that of knowing the plate scale of your image. Plate scale is a term that originates from when photographic plates were used as the main imaging device and is often given in arcsec/mm. For a CCD user, however, a more convenient unit for the plate scale is arcsec/pixel. Clearly the conversion from one to the other is simple.

The focal ratio of a telescope is given by

$$f/ = \frac{\text{focal length of primary mirror}}{\text{primary mirror diameter}},$$

where both values are in the same units and "primary mirror" would be replaced by "primary objective lens" for a refractor. Taking the focal length of the primary (f) in mm and the CCD pixel size (μ) in microns, we can calculate the CCD plate scale as

$$P = \frac{206\,265 \times \mu}{1000 \times f} \quad \text{(arcsec/pixel)},$$

where 206 265 is the number of arcseconds in 1 radian and 1000 is the conversion factor between millimeters and microns.

For a 1-m telescope of $f/ = 7.5$, the focal length (f) of the primary would be 7500 mm. If we were to use a Loral CCD with 15-micron pixels as an imager, the above expression would yield an image scale on the CCD of 0.41 arcsec/pixel. This image scale is usually quite a good value for direct imaging applications for which the seeing is near 1 or so arcseconds.

There are times, however, when the above expression for the plate scale of a CCD may not provide an accurate value. This could occur if there are additional optics within the instrument that change the final f-ratio in some unknown manner. Under these conditions, or simply as an exercise to check the above calculation, one can determine the CCD plate scale observationally. Using a few CCD images of close optical double stars with known separations (e.g., the Washington Double Star Catalog – http://ad.usno.navy.mil/wds/), measurement of the center positions of the two stars and application of a bit of plane geometry will allow an accurate determination of the pixel-to-pixel spacing, and hence the CCD plate scale. This same procedure also allows one to measure the rotation of the CCD with respect to the cardinal directions using known binary star position angles.

4.2 Flat fielding

To CCD experts, the term "flat field" can cause shivers to run up and down their spine. For the novice, it is just another term to add to the lexicon of CCD jargon. If you are in the latter category, don't be put off by these statements but you might want to take a minute and enjoy your thought of "How can a flat field be such a big deal?" In principle, obtaining flat field images and flat fielding a CCD image are conceptually easy to understand, but in practice the reality that CCDs are not perfect imaging devices sets in.

The idea of a flat field image is simple. Within the CCD, each pixel has a slightly different gain or QE value when compared with its neighbors. In order to flatten the relative response for each pixel to the incoming radiation, a flat field image is obtained and used to perform this calibration. Ideally, a flat field image would consist of uniform illumination of every pixel by a light source of identical spectral response to that of your object frames. That is, you want the flat field image to be spectrally and spatially flat. Sounds easy, doesn't it? Once a flat field image is obtained, one then simply divides each object frame by it and voilà: instant removal of pixel-to-pixel variations.

Before talking about the details of the flat fielding process and why it is not so easy, let us look at the various methods devised to obtain flat field exposures with a CCD. All of these methods involve a light source that is brighter than any astronomical image one would observe. This light source provides a CCD calibration image of high signal-to-noise ratio. For imaging applications, one very common procedure used to obtain a flat field image is to illuminate the inside of the telescope dome (or a screen mounted on the inside of the dome) with a light source, point the telescope at the bright spot on the dome, and take a number of relatively short exposures so as not to saturate the CCD. Since the pixels within the array have different responses to different colors of light, flat field images need to be obtained through each filter that is to be used for your object observations. As with bias frames discussed in the last chapter, five to ten or more flats exposed in each filter should be obtained and averaged together to form a final or master flat field, which can then be used for calibration of the CCD. Other methods of obtaining a CCD flat field image include taking CCD exposures of the dawn or dusk sky or obtaining spatially offset images of the dark night sky; these can then be median filtered to remove any stars that may be present (Tyson, 1990; Gilliland, 1992; Massey & Jacoby, 1992; Tobin, 1993).

To allow the best possible flat field images to be obtained, many observatories have mounted a flat field screen on the inside of each dome and painted this screen with special paints (Massey & Jacoby, 1992) that help to reflect all incident wavelengths of light as uniformly as possible. In addition, most instrument user manuals distributed by observatories discuss the various methods of obtaining flat field exposures that seem to work best for their CCD systems. Illumination of dome flat field screens has been done by many methods, from a normal 35-mm slide projector, to special "hot filament" quartz lamps, to various combinations of lamps of different color temperature and intensity mounted like headlamps on the front of the telescope itself. Flat fields obtained by observation of an illuminated dome or dome screen are referred to as dome flats, while observations of the twilight or night sky are called sky flats.

A new generation of wide-field imagers and fast focal length telescopes presents some problems for the normal "dome" screen approach to flat fielding. To achieve large-scale, uniform flat fields Zhou *et al.* (2004) have developed a method by which an isotropic diffuser is placed in front of the telescope and illuminated by reflected light from the dome screen. They claim to obtain flat fields with a measurement of the detector inhomogeneities as good as supersky flats over a 1° field of view. Shi and Wang (2004) discuss flat fielding for a wide field multi-fiber spectroscopic telescope. They use a combination of fiber lamp flat fields and offset sky flats to calibrate the pixel-to-pixel variations.

CCD imaging and photometric applications use dome or sky flats as a means of calibrating out pixel-to-pixel variations. For spectroscopic applications, flat fields are obtained via illumination of the spectrograph slit with a quartz or other high intensity projector lamp housed in an integrating sphere (Wagner, 1992). The output light from the sphere attempts to illuminate the slit, and thus the grating of the spectrograph, in a similar manner to that of the astronomical object of interest. This type of flat field image is called a projector flat. While the main role of a flat field image is to remove pixel-to-pixel variations within the CCD, these calibration images will also compensate for any image vignetting and for time-varying dust accumulation, which may occur on the dewar window and/or filters within the optical path.

Well, so far so good. So what is the big deal about flat field exposures? The problems associated with flat field images and why they are a topic discussed in hushed tones in back rooms may still not be obvious to the reader. There are two major concerns. One is that uniform illumination of every CCD pixel (spatially flat) to one part in a thousand is often needed but in practice is very hard to achieve. Second, QE variations within the CCD pixels are wavelength dependent. This wavelength dependence means that your flat field image should have the exact wavelength distribution over the band-pass of interest (spectrally flat) as that of each and every object frame you wish to calibrate. Quartz lamps and twilight skies are not very similar at all in color temperature (i.e., spectral shape) to that of a dark nighttime sky filled with stars and galaxies.[1] Sky flats obtained of the dark nighttime sky would seem to be our savior here, but these types of flat fields require long exposures to get the needed signal-to-noise ratio and multiple exposures with spatial offsets to allow digital filtering (e.g., median) to be applied in order to remove the stars. In addition, the time needed to obtain (nighttime) sky flats is likely not available to the observer who generally receives only a limited stay at the telescope. Thus, whereas very good calibration data lead to very good final results, the fact is that current policies of telescope scheduling often mean that we must somehow compromise the time used for calibration images with that used to collect the astronomical data of interest. Modern telescopes often observe in queue mode or service mode thereby removing the "at the telescope" interaction of the observer whose data are being collected from the data collection process itself. Often the calibration frames desired are not what is obtained.

---

[1] One good sky region for twilight flats has been determined to be an area 13° east of zenith just after sunset (Chromey & Hasselbacher, 1996).

Flat fielding satellite CCD imagers (such as those on HST) and space misions (such as Cassini) are often hard to achieve in practice. Laboratory flat fields taken prior to launch are often used as defaults for science observations taken in orbit. Defocused or scanned observations of the bright Earth or Moon are often used for these. Dithered observations of a star field can be used as well in a slightly different way. Multiple observations of the same (assumed constant) stars as they fall on different pixels are used to determine the relative changes in brightness and thus map out low frequency variations in the CCD. An example of such a program is discussed in Mack *et al.* (2002).

Within the above detailed constraints on a flat field image, it is probably the case that obtaining a perfect, color-corrected flat field is an impossibility. But all is not lost. Many observational projects do not require total perfection of a flat field over all wavelengths or over the entire two-dimensional array. Stellar photometry resulting in differential measurements or on-band/off-band photometry searching for particular emission lines are examples for which one only needs to have good flat field information over small spatial scales on the CCD. However, a project with end results of absolute photometric calibration over large spatial extents (e.g., mapping of the flux distribution within the spiral arms of an extended galaxy) does indeed place stringent limits on flat fielding requirements. For such demanding observational programs, some observers have found that near-perfect flats can be obtained through the use of a combination of dome and sky flats. This procedure combines the better color match and low-spatial frequency information from the dark night sky with the higher signal-to-noise, high spatial frequency information of a dome flat. Experimentation to find the best method of flat fielding for a particular telescope, CCD, and filter combination, as well as for the scientific goals of a specific project, is highly recommended.

A summary of the current best wisdom on flat fields depends on who you talk to and what you are trying to accomplish with your observations. The following advice is one person's view.

What does the term "a good flat field" mean? An answer to that question is: a good flat field allows a measurement to be transformed from its instrumental values into numeric results in a standard system that results in an answer that agrees with other measurements made by other observers. For example, if two observers image the same star, they both observe with a CCD using a V filter, and they each end up with the final result of $V = 14.325$ magnitudes in the Johnson system then, assuming this is an accurate result, one may take this as an indication of the fact that each observer used correct data reduction and analysis procedures (including their flat fielding) for the observations.

The above is one way to answer the question, but it still relies on the fact that observers need to obtain good flat fields. Without them, near perfect agreement of final results is unlikely. While the ideal flat field would uniformly illuminate the CCD such that every pixel would receive equal amounts of light in each color of interest, this perfect image is generally not produced with dome screens, the twilight sky, or projector lamps within spectrographs. This is because good flat field images are all about color terms. That is, the twilight sky is not the same color as the nighttime sky, neither of which are the same color as a dome flat. If you are observing red objects, you need to worry more about matching the red color in your flats; for blue objects you worry about the blue nature of your flats. Issues to consider include the fact that if the Moon is present, the sky is bluer then when the Moon is absent, dome flats are generally reddish due to their illumination by a quartz lamp of relatively low filament temperature, and so on. Thus, just as in photometric color transformations, the color terms in flat fields are all important. One needs to have a flat field that is good, as described above, plus one that also matches the colors of interest to the observations at hand.

Proper techniques for using flat fields as calibration images will be discussed in Section 4.5. Modern CCDs generally have pixels that are very uniform, especially the new generation of thick, front-side devices. Modern thinning processes result in more even thickness across a CCD reaching tolerances of 1-2 microns in some cases. Thus, at some level flat fielding appears to be less critical today but the advances resulting in lower overall noise performance provide a circular argument placing more emphasis on high quality flats. Appendix A offers further reading on this subject and the material presented in Djorgovski (1984), Gudehus (1990), Tyson (1990), and Sterken (1995) is of particular interest concerning flat fielding techniques.

4.3 Calculation of read noise and gain

We have talked about bias frames and flat field images in the text above and now wish to discuss the way in which these two types of calibration data may be used to determine the read noise and gain for a CCD.

Noted above, when we discussed bias frames, was the fact that a histogram of such an image (see Figure 3.8) should produce a Gaussian distribution with a width related to the read noise and the gain of the detector. Furthermore, a similar relation exists for the histogram of a typical flat field image (see Figure 4.1). The mean level in the flat field shown in Figure 4.1

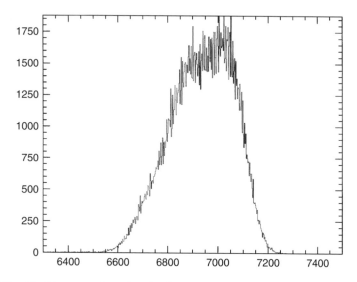

Fig. 4.1. Histogram of a typical flat field image. Note the fairly Gaussian shape of the histrogram and the slight tail extending to lower values. For this R-band image, the filter and dewar window were extremely dusty leading to numerous out of focus "doughnuts" (see Figure 4.4), each producing lower than average data values.

is $\bar{F} = 6950$ ADU and its width (assuming it is perfectly Gaussian (Massey & Jacoby, 1992)) will be given by

$$\sigma_{\mathrm{ADU}} = \frac{\sqrt{\bar{F} \cdot \mathrm{Gain}}}{\mathrm{Gain}}.$$

We have made the assumption in this formulation that the Poisson noise of the flat field photons themselves is much greater than the read noise. This is not unreasonable at all given the low values of read noise in present day CCDs.

Let us now look at how bias frames and flat field images can be used to determine the important CCD properties of read noise and gain. Using two bias frames and two equal flat field images, designated 1 and 2, we can proceed as follows. Determine the mean pixel value within each image.[1] We will call the mean values of the two bias frames $\bar{B}_1$ and $\bar{B}_2$ and likewise $\bar{F}_1$ and $\bar{F}_2$ will be the corresponding values for the two flats. Next, create two difference images ($B_1 - B_2$ and $F_1 - F_2$) and measure the standard deviation

---

[1] Be careful here not to use edge rows or columns, which might have very large or small values due to CCD readout properties such as amplifier turn on/off (which can cause spikes). Also, do not include overscan regions in the determination of the mean values.

of these image differences: $\sigma_{B_1-B_2}$ and $\sigma_{F_1-F_2}$. Having done that, the gain of your CCD can be determined from the following:

$$\text{Gain} = \frac{(\bar{F}_1 + \bar{F}_2) - (\bar{B}_1 + \bar{B}_2)}{\sigma^2_{F_1-F_2} - \sigma^2_{B_1-B_2}},$$

and the read noise can be obtained from

$$\text{Read noise} = \frac{\text{Gain} \cdot \sigma_{B_1-B_2}}{\sqrt{2}}.$$

4.4 Signal-to-noise ratio

Finally we come to one of the most important sections in this book, the calculation of the signal-to-noise (S/N) ratio for observations made with a CCD.

Almost every article written that contains data obtained with a CCD and essentially every observatory user manual about CCDs contains some version of an equation used for calculation of the S/N of a measurement. S/N values quoted in research papers, for example, do indeed give the reader a feel for the level of goodness of the observation (i.e., a S/N of 100 is probably good while a S/N of 3 is not), but rarely do the authors discuss how they performed such a calculation.

The equation for the S/N of a measurement made with a CCD is given by

$$\frac{S}{N} = \frac{N_*}{\sqrt{N_* + n_{\text{pix}}(N_S + N_D + N_R^2)}},$$

unofficially named the "CCD Equation" (Mortara & Fowler, 1981). Various formulations of this equation have been produced (e.g., Newberry (1991) and Gullixson (1992)), all of which yield the same answers of course, if used properly. The "signal" term in the above equation, N_*, is the total number of photons[1] (signal) collected from the object of interest. N_* may be from one pixel (if determining the S/N of a single pixel as sometimes is done for a background measurement), or N_* may be from several pixels, such as all of those contained within a stellar profile (if determining the S/N for the

---

[1] Throughout this book, we have and will continue to use the terms photons and electrons interchangeably when considering the charge collected by a CCD. In optical observations, every photon that is collected within a pixel produces a photoelectron; thus they are indeed equivalent. When talking about observations, it seems logical to talk about star or sky photons, but for dark current or read noise discussions, the number of electrons measured seems more useful.

measurement of a star), or N_* may even be from say a rectangular area of X by Y pixels (if determining the S/N in a portion of the continuum of a spectrum).

The "noise" terms in the above equation are the square roots of N_*, plus n_{pix} (the number of pixels under consideration for the S/N calculation) times the contributions from N_S (the total number of photons per pixel from the background or sky), N_D (the total number of dark current electrons per pixel), and N_R^2 (the total number of electrons per pixel resulting from the read noise.[1]

For those interested in more details of each of these noise terms, how they are derived, and why each appears in the CCD Equation, see Merline & Howell (1995). In our short treatise, we will remark on some of the highlights of that paper and present an improved version of the CCD Equation. However, let's first make sense out of the equation just presented.

For sources of noise that behave under the auspices of Poisson statistics (which includes photon noise from the source itself), we know that for a signal level of N, the associated 1 sigma error (1σ) is given by $\sqrt{N}$. The above equation for the S/N of a given CCD measurement of a source can thus be seen to be simply the signal (N_*) divided by the summation of a number of Poisson noise terms. The n_{pix} term is used to apply each noise term on a per pixel basis to all of the pixels involved in the S/N measurement and the N_R term is squared since this noise source behaves as shot noise, rather than being Poisson-like (Mortara & Fowler, 1981). We can also see from the above equation that if the total noise for a given measurement $\sqrt{N_* + n_{pix}(N_S + N_D + N_R^2)}$ is dominated by the first noise term, N_* (i.e., the noise contribution from the source itself), then the CCD Equation becomes

$$\frac{S}{N} = \frac{N_*}{\sqrt{N_*}} = \sqrt{N_*},$$

yielding the expected result for a measurement of a single Poisson behaved value.

This last result is useful as a method of defining what is meant by a "bright" source and a "faint" source. As a working definition, we will use the term bright source to mean a case for which the S/N errors are dominated by the source itself (i.e., S/N $\sim \sqrt{N_*}$), and we will take a faint source to be the case in which the other error terms are of equal or greater significance compared with N_*, and therefore the complete error equation (i.e., the CCD Equation) is needed.

---

[1] Note that this noise source is not a Poisson noise source but a shot noise; therefore it enters into the noise calculation as the value itself, not the square root of the value as Poisson noise sources do.

The CCD Equation above provides the formulation for a S/N calculation given typical conditions and a well-behaved CCD. For some CCD observations, particularly those that have high background levels, faint sources of interest, poor spatial sampling, or large gain values, a more complete version of the error analysis is required. We can write the complete CCD Equation (Merline & Howell, 1995) as

$$\frac{S}{N} = \frac{N_*}{\sqrt{N_* + n_{\text{pix}} \left(1 + \frac{n_{\text{pix}}}{n_B}\right)\left(N_S + N_D + N_R^2 + G^2\sigma_f^2\right)}}.$$

This form of the S/N equation is essentially the same as that given above, but two additional terms have been added. The first term, $(1 + n_{\text{pix}}/n_B)$, provides a measure of the noise incurred as a result of any error introduced in the estimation of the background level on the CCD image. The term n_B is the total number of background pixels used to estimate the mean background (sky) level. One can see that small values of n_B will introduce the largest error as they will provide a poor estimate of the mean level of the background distribution. Thus, very large values of n_B are to be preferred but clearly some trade-off must be made between providing a good estimate of the mean background level and the use of pixels from areas on the CCD image that are far from the source of interest or possibly of a different character.

The second new term added into the complete S/N equation accounts for the error introduced by the digitization noise within the A/D converter. From our discussion of the digitization noise in Chapter 3, we noted that the error introduced by this process can be considerable if the CCD gain has a large value. In this term, $G^2\sigma_f^2$, G is the gain of the CCD (in electrons/ADU) and σ_f is an estimate of the 1 sigma error introduced within the A/D converter[1] and has a value of approximately 0.289 (Merline & Howell, 1995).

In practice for most CCD systems in use and for most observational projects, the two additional terms in the complete S/N equation are often very small error contributors and can be ignored. In the instances for which they become important – for example, cases in which the CCD gain has a high value (e.g., 100 electrons/ADU), the background level can only be estimated with a few pixels (e.g., less than 200), or the CCD data are of poor pixel

---

[1] The parameter σ_f^2 and its value depend on the actual internal electrical workings of a given A/D converter. We assume here that for a charge level that is half way in between two output ADU steps (that is, 1/2 of a gain step), there is an equal chance that it will be assigned to the lower or to the higher ADU value when converted to a digital number. See Merline & Howell (1995) for further details.

sampling (see Section 5.9) – ignoring these additional error terms will lead to an overestimation of the S/N value obtained from the CCD data.

Let us work through an example of a S/N calculation given the following conditions. A 300-second observation is made of an astronomical source with a CCD detector attached to a 1-m telescope. The CCD is a Thomson 1024×1024 device with 19-micron pixels and it happens that in this example the telescope has a fast f-ratio such that the plate scale is 2.6 arcsec/pixel.[1] For this particular CCD, the read noise is 5 electrons/pixel/read, the dark current is 22 electrons/pixel/hour, and the gain (G) is 5 electrons/ADU. Using 200 background pixels surrounding our object of interest from which to estimate the mean background sky level, we take a mean value for N_B of 620 ADU/pixel. We will further assume here (for simplicity) that the CCD image scale is such that our source of interest falls completely within 1 pixel (good seeing!) and that after background subtraction (see Section 5.1), we find a value for N_* of 24 013 ADU. Ignoring the two additional minor error terms discussed above (as the gain is very small and $n_B = 200$ is quite sufficient in this case), we can write the CCD Equation as

$$\frac{S}{N} = \frac{24\,013(\text{ADU}) \cdot G}{\sqrt{24\,013(\text{ADU}) \cdot G + (1) \cdot (620(\text{ADU}) \cdot G + 1.8 + 5^2(e^-))}}.$$

Note that all of the values used in the calculation of the S/N are in electrons, *not* in ADUs. The S/N value calculated for this example is ~342, a very high S/N. With such a good S/N measurement, one might suspect that this is a bright source. If we compare $\sqrt{N_*}$ with all the remaining error terms, we see that indeed this measurement has its noise properties dominated by the Poisson noise from the source itself and the expression $\text{S/N} \sim \sqrt{N_*} = 346$ works well here.

While the S/N of a measurement is a useful number to know, at times we would prefer to quote a standard error for the measurement as well. Using the fact that $\text{S/N} = 1/\sigma$, where σ is the standard deviation of the measurement, we can write

$$\sigma_{\text{magnitudes}} = \frac{1.0857\sqrt{N_* + p}}{N_*}.$$

In this expression, p is equal to $n_{\text{pix}}(1 + n_{\text{pix}}/n_B)(N_S + N_D + N_R^2 + G^2\sigma_f^2)$, the same assumptions apply concerning the two "extra" error terms, and the value of 1.0857 is the correction term between an error in flux (electrons) and that same error in magnitudes (Howell, 1993). We again see that if the Poisson error of N_* itself dominates, the term p can be ignored and this equation

---

[1] Using the results from Section 4.1, what would be the f-ratio of this telescope?

reduces to that expected for a 1σ error estimate in the limiting case of a bright object.

Additionally, one may be interested in a prediction of the S/N value likely to be obtained for a given CCD system and integration time. N_* is really $N \cdot t$, where N is the count rate in electrons (photons) per second (for the source of interest) and t is the CCD integration time. Noting that the integration time is implicit in the other quantities as well, we can write the following (Massey, 1990):

$$\frac{S}{N} = \frac{Nt}{\sqrt{Nt + n_{\text{pix}}\left(N_S t + N_D t + N_R^2\right)}},$$

in which we have again ignored the two minor error terms. This equation illustrates a valuable rule of thumb concerning the S/N of an observation: $S/N \propto \sqrt{t}$, not to t itself. Solving the above expression for t we find

$$t = \frac{-B + (B^2 - 4AC)^{1/2}}{2A},$$

where $A = N^2$, $B = -(S/N)^2(N + n_{\text{pix}}(N_S + N_D))$, and $C = -(S/N)^2\, n_{\text{pix}} N_R^2$. Most instrument guides available at major observatories provide tables that list the count rate expected for an ideal star (usually 10th magnitude and of 0 color index) within each filter and CCD combination in use at each telescope. Similar tables provide the same type of information for the observatory spectrographs as well. The tabulated numeric values, based on actual CCD observations, allow the user, via magnitude, seeing, or filter width, to scale the numbers to a specific observation and predict the S/N expected as a function of integration time.

4.5 Basic CCD data reduction

The process of standard CCD image reduction makes use of a basic set of images that form the core of the calibration and reduction process (Gullixson, 1992). The types of images used are essentially the same (although possibly generated by different means) in imaging, photometric, and spectroscopic applications. This basic set of images consists of three calibration frames – bias, dark, and flat field – and the data frames of the object(s) of interest. Table 4.1 provides a brief description of each image type and Figures 4.2–4.5

Table 4.1. *Types of CCD images*

CCD Image Type	Image Description
Bias	This type of CCD image has an exposure time of zero seconds. The shutter remains closed and the CCD is simply read out. The purpose of a bias or zero frame is to allow the user to determine the underlying noise level within each data frame. The bias value in a CCD image is usually a low spatial frequency variation throughout the array, caused by the CCD on-chip amplifiers. This variation should remain constant with time. The rms value of the bias level is the CCD read noise. A bias frame contains both the DC offset level (overscan) and the variations on that level. The nature of the bias variations for a given CCD are usually column-wise variations, but may also have small row-wise components as well. Thus, a 2-D, pixel-by-pixel subtraction is often required. A single bias frame will not sample these variations well in a statistical fashion, so an average bias image of 10 or more single bias frames is recommended.
	★★★★★
Dark	CCD dark frames are images taken with the shutter closed but for some time period, usually equal to that of your object frames. That is, if one is planning to dark correct a 45 second exposure, a 45 second dark frame would be obtained. Longer dark frames can often be avoided using the assumption that the dark current increases linearly with time and a simple scaling can be applied. However, this is not always true. Dark frames are a method by which the thermal noise (dark current) in a CCD can be measured. They also can give you information about bad or "hot" pixels that exist as well as provide an estimate of the rate of cosmic ray strikes at your observing site. Observatory class CCD cameras are usually cooled with LN2 to temperatures at which the dark current is essentially zero. Many of these systems therefore do not require the use of dark exposure CCD frames in the calibration process. Thermoelectrically cooled systems are not cooled to low enough temperatures such that one may ignore the dark current. In addition, these less expensive models often have poor temperature stability allowing the dark current to wander a bit with time. Multiple darks

Table 4.1. *(cont.)*

CCD Image Type	Image Description
	averaged together are the best way to produce the final dark calibration frame. Note that if dark frames are used, the bias level of the CCD is present in them as well, and therefore separate bias frames are not needed.
Flat Field	Flat field exposures are used to correct for pixel-to-pixel variations in the CCD response as well as any nonuniform illumination of the detector itself. Flat fields expose the CCD to light from either a dome screen, the twilight sky, the nighttime sky, or a projector lamp in an attempt to provide a high S/N, uniformly illuminated calibration image. For narrow-band imaging, flats are very helpful in removing fringing, which may occur in object frames. Flat field calibration frames are needed for each color, wavelength region, or different instrumental setup used in which object frames are to be taken. A good flat should remain constant to about 1%, with 2% or larger changes being indicators of a possible problem. As with the other calibration frames, at least 5 or more flat fields should be taken and averaged to produce the final flat used for image calibration.
	★ ★ ★ ★ ★
Object	These are the frames containing the astronomical objects of interest. They are of some exposure length from 1 second or less up to many hours, varying for reasons of type of science, brightness of object, desired temporal sampling, etc. Within each object image pixel is contained contributions from the object and/or sky, read noise, thermally generated electrons, and possibly contributions from cosmic rays. Each pixel responds similarly but not exactly to the incident light, so nonuniformities must be removed. All of the noise and spatial factors are correctable to very low levels via standard CCD reductions as described in the text.

show examples of typical bias, dark, flat field, and object CCD images. Note that a CCD dark frame contains not only information on the level and extent of the dark current but also includes bias level information.

The use of the basic set of calibration images in the reduction of CCD object frames is as follows. First, subtract a mean bias frame (or dark frame

Fig. 4.2. Shown is a typical CCD bias frame. The histrogram of this image was shown in Figure 3.8. Note the overall uniform structure of the bias frame.

if needed[1]) from your object frame. Then, divide the resulting image by a (bias subtracted) mean flat field image. That's all there is to it! These two simple steps have corrected your object frame for bias level, dark current (if needed), and nonuniformity within each image pixel. During the analysis of your object frames, it is likely that the background or sky contribution to the image will need to be removed or accounted for in some manner. This correction for the background sky level in your image frames is performed

---

[1] The need for dark frames instead of simply bias frames depends entirely on the level of dark current expected during an integration or the stability of the dark current from integration to integration. The first situation depends on the operating temperature of the CCD. LN2 systems have essentially zero dark current, and thus bias frames are all that is needed. Inexpensive and thermoelectrically cooled CCD systems fall into the category of generally always needing dark frames as part of the calibration process.

Fig. 4.3. Shown is a typical CCD dark frame. This figure shows a dark frame for a Kodak CCD operating in MPP mode and thermoelectrically cooled. Notice the nonuniform dark level across the CCD, being darker (greater ADU values) on the top. Also notice the two prominent partial columns with higher dark counts, which extend from the top toward the middle of the CCD frame. These are likely to be column defects in the CCD that occurred during manufacture, but with proper dark subtraction they are of little consequence. The continuation of the figure shows the histogram of the dark frame. Most of the dark current in this 180 second exposure is uniformly distributed near a mean value of 180 ADU with a secondary maximum near 350 ADU. The secondary maximum represents a small number of CCD pixels that have nearly twice the dark current of the rest, again most likely due to defects in the silicon lattice. As long as these increased dark current pixels remain constant, they are easily removed during image calibration.

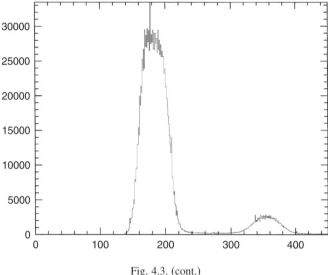

Fig. 4.3. (cont.)

as part of each specific analysis step using "sky" regions in the object frame itself and is not removed or corrected for with some sort of separate "sky" frame. In equational form, the calibration process can be written as

$$\text{Final Reduced Object Frame} = \frac{\text{Raw Object Frame} - \text{Bias Frame}}{\text{Flat Field Frame}},$$

where, again, the flat field image has already been bias subtracted and the bias frame would be replaced by a dark frame when appropriate.

4.6 CCD imaging

This section details issues related to the application of using CCDs to produce images of an extended area of the sky. Examples of this type of CCD observation are multi-color photometry of star clusters, galaxy imaging to isolate star-forming regions within spiral arm structures, deep wide-field searches for quasars, and extended low surface brightness mapping of diffuse nebulae. Use of the areal nature of a CCD introduces some additional issues related to the calibration procedures and the overall cosmetic appearance as any spurious spatial effects will have implications on the output result. We briefly discuss here a few new items that are of moderate concern in two-dimensional imaging and then move on to the topic of wide-field imaging with CCD mosaic cameras.

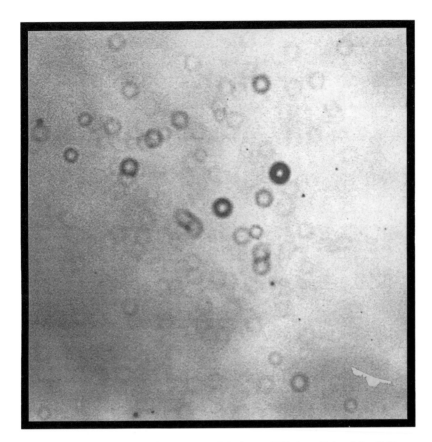

Fig. 4.4. Shown is a typical CCD flat field image. This is an R-band flat field image for a 1024 × 1024 Loral CCD. The numerous "doughnuts" are out of focus dust specks present on the dewar window and the filter. The varying brightness level and structures are common in flat field images. As seen in the histogram of this image (Figure 4.1) this flat field has a mean level near 6950 ADU, with an approximate dispersion of (FWHM) 400 ADU.

4.6.1 CCD fringing and other cosmetic effects

We mentioned earlier that observations of monochromatic (or nearly so) light can cause a pattern of fringes to occur on a CCD image. These fringes, which are essentially Newton's rings, are caused by interference between light waves that reflect within the CCD or long wavelength light that passes through the array and reflects back into the array. Fringing may occur for CCD observations in the red part of the optical spectrum, when narrow-band filters are used, or if observations are made of a spectral regime (e.g., the I-band) that contains strong narrow emission lines. For a given fringe cause (e.g., a specific wavelength set of emission lines) the fringe pattern on the CCD remains constant. Figure 4.6 shows a Gemini North GMOS image obtained in a z' filter (central wavelength is near 8800 Å) on a photometric night with no moon but plenty of OH emission. The GMOS detector consists of three EEV red 13.5 micron 6144 × 4608 CCDs placed next to each other

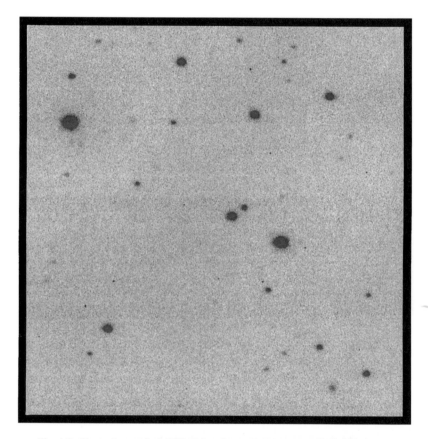

Fig. 4.5. Shown is a typical CCD object frame showing a star field. This image has been properly reduced using bias frame subtraction and division by a flat field image. Note how the background is of a uniform level and distribution; all pixel-to-pixel nonuniformities have been removed in the reduction process. The stars are shown as black in this image and represent R magnitudes of 15th (brightest) to 20th (faintest). The histogram shown in the remainder of the figure is typical for a CCD object frame after reduction. The large grouping of output values on the left (values less than about 125 ADU) are an approximate Gaussian distribution of the background sky. The remaining histogram values (up to 1500 ADU) are the pixels that contain signal levels above the background (i.e., the pixels within the stars themselves!).

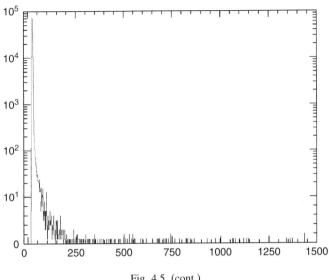

Fig. 4.5. (cont.)

vertically. The frame on the left shows a typical CCD fringe pattern caused by the night sky emission lines while the frame on the right has been defringed. Figure 4.7 presents line plots across typical fringing i' and z' GMOS frames. The typical level of fringing is near ±0.7% in i' and ±2.5% in z'.

The troubling aspect with fringing in CCD data is that it is often the case that the fringe pattern does not occur in the flat field frames (flats contain no emission lines!) or the level of fringing is highly variable throughout the night. Without a pattern match between the flats and the image data, fringe removal will not robustly occur during image calibration, and residual fringes will remain in the final object images. One of the major causes of CCD fringing is the night sky emission lines, which occur in the Earth's upper atmosphere (Pecker, 1970). These night sky lines are mainly attributed to OH transitions in the atmosphere, which are powered by (UV) sunlight during the day. Since they are forbidden transitions they have long decay lifetimes and are very narrow spectrally. In addition, due to upper atmosphere motions, OH concentrations, and their long decay times, these emission lines are highly variable in time and strength, even within a given night. Dealing with fringes that occur in CCD data can be a difficult problem but one for which solutions exist (Broadfoot & Kendall, 1968; Wagner, 1992). Observations with newly defined moderate-band filters that lie between the OH transitions is one such example.

Additionally, cosmetic effects such as bad pixels, hot pixels (LEDs), stuck bits, or dead columns can be present and can mar a CCD image. Not only do these

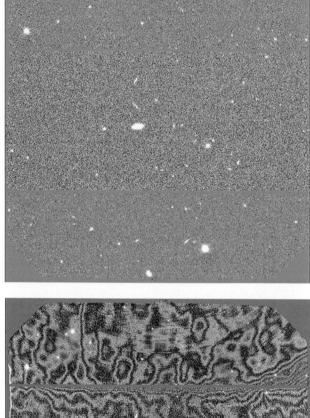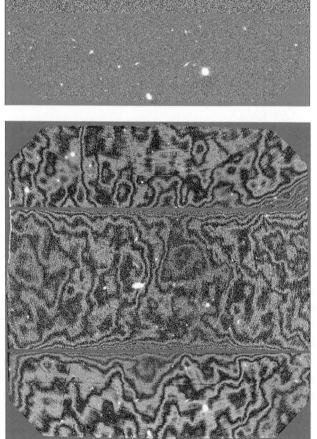

Fig. 4.6. Gemini North GMOS CCD fringe frame (left) and reduced, defringed, frame (right). The night was photometric and near new moon but had OH emission present. Notice that the 1–2% fringes can cross over objects of interest but are mostly fully removed during reduction.

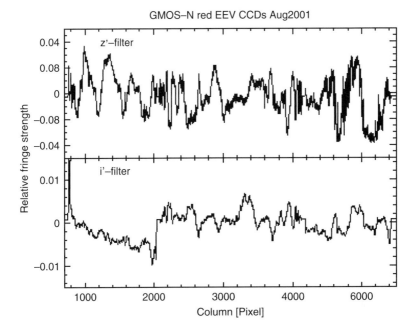

Fig. 4.7. Line plots across the unprocessed GMOS i' and z' images. The plots have been normalized such that the mean image level is zero and the fringe level can be seen to be both positive and negative deviations from this level. The i' fringing is about 0.7% while the z' fringing is near 2.5%.

flaws spoil the beauty of the two-dimensional data, they can cause problems during calibration and analysis by hindering software processes and not allowing correct flux estimates to be made for the pixels that they affect. Procedures for the identification and removal of, or correction for these types of problems can be applied during image calibration and reduction. They are specialized reduction tasks, which depend on the desired output science goals and generally are specific to a particular CCD, instrument, or type of observation being made. Most observatories provide solutions to such fixed flaws. One example is a bad pixel map, which consists of an "image" of 0s and 1s with 0s at the locations of bad columns or other regions of bad pixels. These maps are used by software in the reduction process to eliminate and fix offending CCD problems. A complete discussion of all of these topics lies beyond our space limitations but the interested reader will find discussions of such corrections in Djorgovski (1984), Janesick *et al.* (1987a), Gilliland (1992), Gullixson (1992), Massey & Jacoby (1992), and numerous specific instrument manuals and reference papers concerning the finer points of specific CCD related issues (see Appendix A).

4.6.2 Tip-tilt corrections

The Earth's atmosphere causes a blurring of an image and thus a reduction in image quality during an observation. A solution that often eliminates nearly 70–80% of this effect is the use of adaptive optics to perform low order tip-tilt corrections. Mechanical tip-tilt systems exist today at many observatories and consist of a guide star sensor of some type (avalanche photodiodes (APDs) or a small CCD) and a small optical mirror that can tip and tilt rapidly. The sensor receives light from a bright guide star in the field of view (or a laser guide star) during an observation and the quality (mainly the x,y position) of the image observed by the sensor is assessed. A fast feedback is established by which any movement in the guide star is measured and a correction tip-tilt signal is sent to the moveable mirror.

Systems of this type have small fields of view ($\sim$2–4 arcminutes) and can only work well if a bright guide star is present of if the telescope is equipped with a laser beacon. Orthogonal transfer CCDs were developed to provide nonmechanical tip-tilt corrections. The OTCCD camera OPTIC (Tonry *et al.*, 1997) has four guide regions (at the ends of the CCDs) and four associated science regions. Up to four stars that fall in the guide regions are used for tip-tilt correction. These stars are read out fast (typically 10–20 Hz), assessed, and a tip-tilt correction signal is fed back to the science regions of the CCDs during the integration. OTCCDs can shift charge on the array in both x and y directions and use this property to provide fast tip-tilt correction in the science image. This same type of feedback can also simultaneously correct for telescope drive errors and wind shake (see Tonry *et al.*, 1997, Howell *et al.*, 2003).

The new generation of OTCCD, the OTAs (see Figures 2.7 and 2.8), will extend the tip-tilt correction ability. They allow use of any of the individual OTCCDs within the 8×8 array to be used as a guide region. Additionally, the ability to tip-tilt correct an image can be extended to an arbitrarily large field of view as each part of the OTCCD array corrects itself locally. The WIYN observatory is building a one-degree imager that will provide tip-tilt corrections across the entire 1° field of view. The Pan-STARRS project is developing a similar imager that will cover a 3° field (Jacoby *et al.*, 2002, Tonry *et al.*, 2002).

4.6.3 Wide-field CCD imaging

With the advent of large-footprint CCDs and the construction of CCD mosaic arrays containing many chips, wide-field imaging is becoming one of the major applications in astronomy today. One of the major efforts in observational astronomy today is large field of view, multi-color imaging of the sky. Large surveys such as the Sloan digital sky survey (SDSS) and the two-micron all sky survey (2MASS) are complete and their contribution to astronomy has been amazing. New types of objects, large, very complete samples, and follow-up spectroscopy have shown that imaging surveys can provide tremendous new information.

Temporal variation of the objects in the night sky (both known and unknown) is an additional parameter becoming an integral part of modern imaging surveys. At least six very ambitious wide-field imaging projects are well underway to complement the ten or so, 0.5–1.0° field of view imagers already in action. Table 4.2 lists a few examples of modern wide-field CCD imaging cameras available to the astronomer today as well as those planned to be built and on-line in the next decade. Wide-field imagers of even five years ago consisted of four large format CCDs and required minutes for readout and often days for data reduction. Modern wide-field cameras consist of dozens of CCDs, readout very fast (and will get faster with estimates of 2–4 s), and pass through automated software pipelines in a matter of hours. Figures 4.8–4.10 show two currently working large CCD mosaic cameras plus the planned Pan-STARRS OTCCD camera.

MegaCam on the CFHT was the first operational wide-field, megapixel CCD imager (Boulade *et al.*, 1998) starting science operation in 2002. Today, the Large Synoptic Survey Telescope (LSST) project is the most ambitious of the currently planned wide-field imagers. This special purpose imaging telescope will have a camera containing gigapixels of CCD real estate and image an area of nearly ten square degrees at a time. The LSST camera (see Table 4.1) will likely use an array of 1K or 2K CMOS or CCD ASIC devices. ASIC (Application Specific Integrated Circuits) devices are special purpose production circuits made with a number of non-changeable specific modes built directly into the chip. As such, ASIC devices are often higher in efficiency but somewhat limited in expandability for use other than what they were designed for. An example of a common ASIC device is the computer chip residing under the hood of most modern automobiles. The LSST will image the entire sky every few nights and the amount of data produced will run into the petabytes.

Astronomers are beginning to become different types of observers. Virtual on-line databases, such as the National Virtual Observatory (NVO) will soon sponsor the ability for world-wide access to a tremendous amount of data. Preliminary versions of the web tools and software exist today and with many new CCD imagers available and ever larger ones coming along, there promises to be no shortage of data to sift through and extract scientific research projects from. One downside to this type of observational work is

Table 4.2. *Some Present and Planned Large Telescope CCD Imagers (* marks planned imagers)*

Name	Telescope	Field of View	Focal Plane CCDs	Pixel Size (microns)	Pixel Scale ("/pix)	Website
MageCam	CFHT	0.92 sq. degrees	40 − 2048 × 4612 E2V	13.5	0.185	http://www.cfht.hawaii.edu/Instruments/Imaging/Megacam/
QUEST	Palomar 48″ Schmidt	16 sq. degrees	112 − 600 × 2400 Sarnoff	13	∼1.5	http://www.astro.caltech.edu/∼george/pq/
SUPRIME	Subaru	0.24 sq. degrees	10 −2048 ×4096 MIT/LL	15	0.2	http://www.naoj.org/ Observing/Instruments/SCam/
Mosaic	KPNO/CTIO 4-m	0.36 sq. degrees	8 − 2048 × 4096 SITe	15	0.26	http://www.noao.edu/kpno/mosaic
MegaCam	MMT	0.16 sq. degrees	32 − 2048 × 4096 E2V	15	0.087	http://cfa-www.harvard.edu/∼bmcleod/Megacam/
Dark Energy Camera*	CTIO 4-m	2.9 sq. degrees	70 − 2048 × 4096 LBL	15	0.28	http://www.fnal.gov/pub/
OmegaCam*	VST	1.0 sq. degrees	32 − 2048 × 4096 E2V	15	0.21	http://www.eso.org/instruments/omegacam/
One Degree Imager*	WIYN	1.0 sq. degrees	(OTA) 60 − 3840 × 3952 STA/Dalsa	12	0.11	http://www.noao.edu/wiyn/ODI/
Pan-STARRS*	1.8-m	7.0 sq. degrees	(OTA) 60 − 4096 × 4096 MIT/LL	12	0.3	http://pan-starrs.ifa.hawaii.edu/public/index.html
Kepler* (Spacecraft)	1.0-m Schmidt	105 sq. degrees	42 − 2200 × 1024 E2V	27	4	http://www.kepler.arc.nasa.gov/
LSST Camera*	LSST	9.6 sq. degrees	3 Gigapixels LBL	10	0.2	http://www.lsst.org/lsst_home.shtml

Fig. 4.8. Photograph of the CCDs used in the CFHT MegaCam. Forty large format E2V CCDs are used in this camera, which can image a field of view of nearly 1 × 1 degree on the sky.

that the virtual observer will only be able to get the data that were taken and they may or may not suit their needs. So don't stop thinking of your own observational projects or planning to go to a telescope to collect your own data just yet.

Wide-field CCD mosaic imagers provide a tremendous amount of information (and data) to be collected in one exposure. The soon-to-be-operating OmegaCam on the VLT survey telescope (VST), for example, will produce

Fig. 4.9. A view of the Subaru SuPrime CCDs mounted in the camera dewar. This camera images a field of view of $\sim 0.5°$ on a side using ten 2048×4096 MIT/LL CCDs.

over 4200 Mb of data in one exposure! CCD mosaic arrays are pioneering new scientific advances and driving astronomical technology such as read-out, CCD controllers, and data storage. These larger CCD arrays are also being enabled by faster computational ability and increased effort in software and hardware development. Astronomers have been making a transition from

Internal View of Gigapixel Camera Cryostat

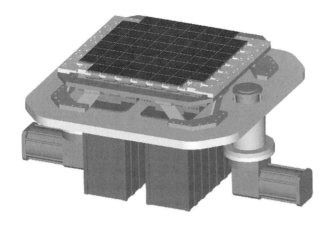

PanSTARRS Gigapixel Camera CoDR · 30
1/29/03

Fig. 4.10. Engineering drawing of one of the four Pan-STARRS imagers currently under construction. This camera will use sixty OTAs to cover a 7 square degree field of view.

single researchers and few night runs at a telescope, to large collaborations that build instruments and telescopes, to the production of extensive non-proprietary databases. Financial constraints and enormous complexity are the prime drivers of this new research model. Physicists went down this road many years ago and we often joke about their papers having less text in the science portion then the two pages that list the 200 authors. Astronomy is going in this direction and the new generation of large, expensive CCD imagers are leading the way.

The efficiency of a large-area survey can be estimated by the metric

$$\epsilon = \Omega D^2 q,$$

where Ω is the solid angle of the field of view, D is the diameter of the telescope, and q is the total throughput quantum efficiency of the instrument assuming that the seeing disk is resolved. One can see that the time needed for completion of a survey to a given brightness limit depends inversely on ϵ.

Using wide-field CCD imagers leads to the inevitable result that new issues of calibration and data reduction must be developed. For example, when the field of view of a large-area CCD (whether a single large CCD with a wide-field of view or an array of chips) approaches $\sim 0.5°$ in size, differential refraction of the images across the field of view of the CCD begins to become important. Color terms therefore propagate across the CCD image and must be corrected for to properly determine and obtain correct photometric information contained in the data.

CCD observations that occur through uniform thin clouds or differential measures essentially independent of clouds are often assumed to be valid, as it is believed that clouds are grey absorbers and that any cloud cover that is present will cover the entire CCD frame. Thus flux corrections (from, say, previous photometric images of the same field) can be applied to the nonphotometric CCD data, making it usable. Large-field CCD imaging can not make such claims. A one or more square degree field of view has a high potential of not being uniformly covered by clouds, leading to unknown flux variations across the CCD image.

Observations of large spatial areas using CCD mosaics also necessitate greater effort and expense in producing larger filters, larger dewar windows, and larger correction optics. Variations of the quality and color dependence of large optical components across the field of view are noticeable and their optical aberrations will cause point-spread function (PSF) changes and other effects over the large areas imaged with wide-field CCDs. Production of large, high quality optical components is a challenge as well. For example, recent estimates for the cost of a single 16 to 20-inch square astronomical quality glass filter are in the range of $50 000–200 000.

The use of large-format CCDs or CCD mosaics on Schmidt telescopes is increasing and such an imager provides a good example of the type of PSF changes that occur across the field of view (see Figure 4.11). Coma and chromatic aberrations are easily seen upon detailed inspection of the PSFs, especially near the corners or for very red or blue objects whose peak flux lies outside of the color range for which the optics were designed. Thus, for wide-field applications, such as that represented in Figures 4.8–4.10, the typical assumption that all the PSFs will be identical at all locations within the field of view must be abandoned.

A more subtle effect to deal with in wide-field imaging is that of the changing image scale between images taken of astronomical objects and those obtained for calibration purposes. For example, a dome flat field image taken for calibration purposes will not have exactly the same image scale per pixel over the entire CCD image as an object frame taken with the same CCD camera, but of an astronomical scene. Also, how one maps the light collected per (non-equal area) pixel in the camera to a stored image in say RA and DEC is a tessellation problem to be solved.

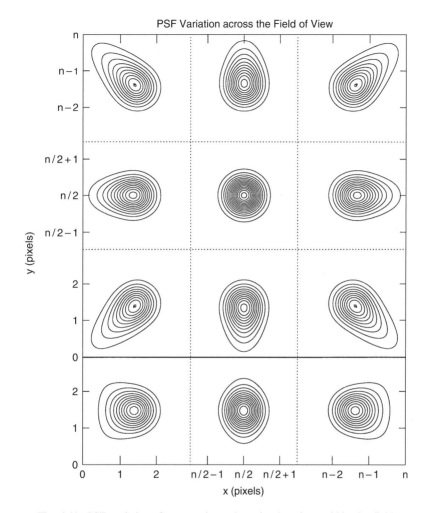

Fig. 4.11. PSF variations for a star imaged at nine locations within the field of view of a large mosaic CCD camera placed at the focal plane of a Schmidt telescope. Only the center of the field has a circular, symmetric PSF while the other positions show extended tails due to optical abberations and chromatic effects. The three PSFs at the bottom of the figure are column sums of the PSFs vertically above them. From Howell *et al.* (1996).

As with previous new advances in CCD imaging, wide-field imaging has issues that must be ironed out. However, this exciting new field of research is still in its infancy and those of you reading this book who are involved in such work are the ones who must help determine the proper data collection and reduction procedures to use.

4.6.4 CCD drift scanning and time-delay integration

The standard method of CCD imaging is to point the telescope at a particular place in the sky, track the telescope at the sidereal rate, and integrate with the detector for a specified amount of time. Once the desired integration time is obtained, the shutter is closed and the CCD is readout. For telescopes incapable of tracking on the sky or to obtain large areal sky coverage without the need for complex CCD mosaics, the techniques of CCD drift scanning and time-delay integration were developed (McGraw, Angel, & Sargent, 1980; Wright & Mackay, 1981).

Drift scanning consists of reading the exposed CCD at a slow rate while simultaneously mechanically moving the CCD itself to avoid image smear. The readout rate and mechanical movement are chosen to provide the desired exposure time. Each imaged object is thus sampled by every pixel in the column thereby being detected with the mean efficiency of all pixels in the column. Nonuniformities between the pixels in a given column are thus eliminated as each final pixel is, in essence, a sum of many short integrations at each pixel within the column. Cross column efficiency differences are still present but the final image can now be corrected with a one-dimensional flat field. Drift scanning also has the additional advantage of providing an ideal color match to background noise contributions, unavailable with dome flats. Very good flat fielding of a traditional image might reach 0.5 or so percent, while a good drift scanned CCD image can be flattened to near 0.1 percent or better (Tyson & Seitzer, 1988). Drift scanning has even been accomplished with IR arrays (Gorjian, Wright, & Mclean, 1997).

Time-delay integration or TDI is a variant on the drift scanning technique. In TDI, the CCD does not move at all but is readout at exactly the sidereal rate. This type of CCD imaging is necessary if astronomical telescopes such as transit instruments (McGraw, Angel, & Sargent, 1980) or liquid mirror telescopes (Gibson, 1991) are to be used. The same flat fielding advantages apply here as in drift scanning but the integration time per object is limited by the size of the CCD (i.e., the time it takes an object to cross the CCD field of view). For a 2048×2048 CCD with 0.7 arcsec pixels, the integration time would be only 96 seconds at the celestial equator. Rescanning the same area could be performed and co-added to previous scans as a method of increasing the exposure time, but time sampling suffers.

TDI is mechanically simple, as nothing moves but the electrons in the CCD. This charge movement has been termed electro-optical tracking. Large sky regions can be surveyed, albeit to shallow magnitude limits, very quickly using TDI. Overhead time costs for TDI only consist of the "ramp up" time, that is, the time needed for the first objects to cross the entire field, and the scan time. Using our same 2048×2048 CCD as in the example above, we find that a 23 arcsec by 3 degree long strip of the sky at the celestial equator can be scanned in about 2–3 minutes compared with the nearly 25 minutes required if pointed observations of equivalent integration are used.

Although drift scanning and TDI are seemingly great solutions to flat fielding issues and offer the collection of large datasets, drift scanning requires the CCD to move during the integration with very precise and repeatable steps. This is quite a mechanical challenge and will increase the cost of such an instrument over that of a simple CCD imager. In addition, both techniques suffer two potential drawbacks (Gibson & Hickson, 1992). Images obtained by drift scanning and TDI techniques have elongated PSFs in the east–west direction. This is due to the fact that the rows of the CCD are shifted discretely while the actual image movement is continuous. We note here that objects seperated by even small declination differences (i.e., one CCD field of view) do not have the same rate of motion. The resulting images are elongated east–west and are a convolution of the seeing with the CCD pixel sampling.

TDI imagery contains an additional distortion in the north–south direction due to the cuvature of an object's path across the face of the CCD (if imaging away from the celestial equator). This type of distortion is usually avoided in drift scan applications as the telescope and CCD tracking are designed to eliminate this image smearing. This sort of mechanical correction can not be applied to TDI imaging.

These image deformations have been studied in detail (Gibson & Hickson, 1992) and are seen to increase in magnitude for larger format CCDs or declinations further from the celestial equator. For example, at a declination of $\pm 30°$, a 1 arcsec per pixel CCD will show an image smear of about 6 pixels. One solution to this large image smear is to continuously reorient the CCD through rotations and translations, such that imaging scans are conducted along great circles on the sky rather than a polar circle or at constant declination. Such a mechanically complex device has been built and used for drift scanning on the 1-m Las Campanas telescope (Zaritsky, Shectman, & Bredthauer, 1996). Another solution is the development of a multilens optical corrector that compensates for the image distortions by tilting and decentering the component lenses (Hickson & Richardson, 1998).

A few telescopes have made good use of the technique of drift scanning or TDI, providing very good astronomical results. Probably the first such project was the Spacewatch telescope (Gehrels *et al.*, 1986) built to discover and provide astrometry for small bodies within the solar system. Other notable examples are the 2-m transit telescope previously operated on Kitt Peak (McGraw, Angel, & Sargent, 1980) and a 2.7-m liquid mirror telescope currently running at the University of British Columbia (Hickson *et al.*, 1994). This latter telescope contains a rotating mercury mirror and images a 21-arcminute strip of the zenith with an effective integration time of 130 seconds. Using TDI, a typical integration with this liquid mirror telescope reaches near 21st magnitude in R and continuous readout of the CCD produces about 2 Gb of data per night.

Present-day examples of telescopes employing drift scanning and TDI techniques are the QUEST telescope (Sabby, Coppi, & Oemler, 1998), the Palomar QUEST imager (see Table 4.2) and the Sloan digital sky survey (Gunn *et al.*, 1998). The QUasar Equatorial Survey Team (QUEST) telescope is a 1-m Schmidt telescope that will provide UBV photometry of nearly 4000 square degrees of the sky to a limiting magnitude of near 19. The focal plane will contain sixteen 2048×2048 Loral CCDs arranged in a 4×4 array. The telescope is parked and the CCDs are positioned such that the clocking (column) direction is east–west and the readout occurs at the apparent sidereal rate. Each object imaged passes across four CCDs covered, in turn, with a broadband V, U, B, and V filter. The effective integration time (i.e., crossing time) is 140 seconds, providing nearly simultaneous photometry in U, B, and V.

As we have seen above, a problem with drift scanning is that the paths of objects that drift across the imager are not straight and they can cross the wide-field of view with different drift rates. We have discussed a few solutions to these issues, and in the QUEST project (Sabby, Coppi, & Oemler, 1998) we find another. The CCDs are fixed, in groups of four, to long pads lying in the north–south direction. These pads can pivot independently such that they align perpendicular to the direction of the stellar paths. The CCDs are also able to be clocked at different rates, with each being readout at the apparent sidereal rate appropriate for its declination.

The Sloan digital sky survey (SDSS) is a large-format mosaic CCD camera consisting of a photometric array of thirty 2048×2048 SITe CCDs and an astrometric array of twenty four 400×2048 CCDs (see Figure 4.12). The photometric CCDs are arranged in six columns of five CCDs each, providing essentially simultaneous five-color photometry of each image object. The astrometric CCDs are mounted in the focal plane above and below the main array and will be used to provide precise positional information needed for the follow-up multi-fiber spectroscopy. The SDSS uses a 2.5-m telescope located in New Mexico to image one quarter of the entire sky down to a limiting magnitude of near 23.

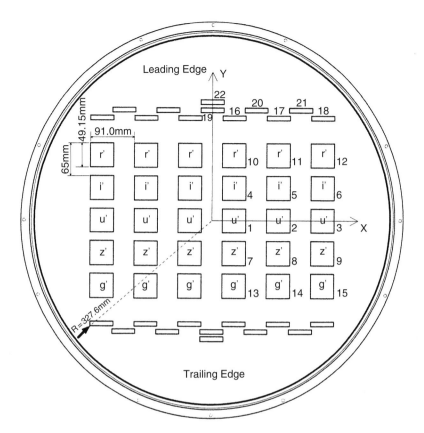

Fig. 4.12. The optical layout within the dewar of the Sloan digital sky survey CCD imager. The right side of the figure labels the CCDs as to their function; 1–15 are photometric CCDs, 16–21 are astrometric CCDs, and 22 (top and bottom) are focus CCDs. The left side gives the dimensions of the array. The labels r'–g' denote the five separate intermediate band filters, each a single piece of glass covering all six horizontal CCDs. The scan direction is upward causing objects to traverse the array from top to bottom. From Gunn *et al.* (1998).

TDI scans along great circles are used by the SDSS to image a region of the sky 2.5° wide. Using five intermediate band filters, covering 3550 Å to 9130 Å, scanning at the sidereal rate provides an effective integration per color of 54 seconds with a time delay of 72 seconds between colors caused by CCD crossing time and chip spacing. Complete details of the SDSS, too lengthy for presentation here, can be found in Gunn *et al.* (1998). Other projects of a similar nature are discussed in Boulade *et al.* (1998), Gunn *et al.* (1998), and Miyazaki *et al.* (1998). The SDSS prime survey is complete and much of the data are already available (see Appendix B).

4.7 Exercises

1. Derive the first two equations of Chapter 4.

2. What focal length (f-ratio) of telescope is best if your observational requirements need a plate scale of about one tenth of an arcsec/pixel? How does your answer depend on the type of CCD used? What is the f-ratio of a typical present-day large reflecting telescope?

3. Design an experiment to obtain a good flat field image, that is, one that will allow you to measure the pixel-to-pixel variations to 1%. How might your experiment differ if you were to make the measurements in the red? In the blue? Of the night sky?

4. What are the differences between a sky flat and a dome flat? Which is easier to obtain? Which provides the better flat field?

5. What are the flat fielding requirements needed for point source photometry? For extended object spectroscopy? For extended object imaging? How might you accomplish each of these?

6. Using the method outlined in Section 4.3, determine the gain and read noise for a CCD you work with.

7. Using the equations presented in Section 4.4, calculate the individual noise contribution per pixel from read noise, sky background, and dark current given the following conditions. You are using an E2V CCD at operating temperature (as described in Table 3.2) and have obtained a 1200 second exposure on a full moon night using a Johnson V filter. (Note: You will have to estimate the sky brightness (see an observatory website for such details), plate scale, and band-pass of your observation.) Plot your results. Which noise source is the greatest? How might you eliminate it?

8. Derive the signal-to-noise equation.

9. Describe an observational setup for which a 15th magnitude galaxy is a bright source. Do the same for a faint source.

10. Work through the example S/N calculation given in Section 4.4.

11. Answer the question posed in the footnote on page 76.

12. Derive the expression for the integration time needed to achieve a specific S/N as given at the end of Section 4.4.

13. Produce a flow chart of a typical reduction procedure for CCD imaging observations. Clearly show which types of calibration images are needed and when they enter into the reduction process.

14. Why do you divide object frames by a flat field calibration image instead of multiplying by it?

15. Look at Figure 4.4. If the doughnuts are due to out-of-focus dust, how might you be able to use their size or shape to tell if that dust was on the dewar window or on a filter high above the CCD dewar?

16. Using information on the wavelengths and strengths of night sky emission lines (see e.g., Broadfoot and Kendall, 1968, Pecker, 1970, or observatory websites), discuss which broad-band Johnson filters are likely to be affected by these lines. How might one design an observational program that uses these filters but lessens the effect of the night sky lines?

17. Using the physical principles of Newton's Rings, quantitatively describe CCD fringing providing a relationship between the CCD thickness and the wavelength observed.

18. Discuss how OTCCDs can provide tip-tilt corrections over an arbitrary field of view. Why can mechanical tip-tilt systems not do this?

19. Using the expression given for the efficiency of a large area survey, calculate the efficiency for each imaging program listed in Table 4.2. How does the LSST project compare to the rest? How do your values compare with a survey using a 4-m f/1.5 telescope able to image onto a 14 × 14 inch photographic plate?

20. Design an observational experiment to map galaxy clusters using a CCD system that operates in drift scanning mode. Discuss the details of observational strategy, integration times, instrument design, and calibration. Where would you locate your survey telescope and why is this important? How does this type of observation program compare with a similar one that uses a conventional point-and-shoot CCD system?

21. Read the description of the Sloan survey given in Gunn *et al.* (1998). Discuss why this survey is important to astronomy. Can you think of any improvements you would make to the methods used if you were designing the survey?

5

Photometry and astrometry

One of the basic astronomical pursuits throughout history has been to determine the amount and temporal nature of the flux emitted by an object as a function of wavelength. This process, termed photometry, forms one of the fundamental branches of astronomy. Photometry is important for all types of objects from planets to stars to galaxies, each with their own intricacies, procedures, and problems. At times, we may be interested in only a single measurement of the flux of some object, while at other times we could want to obtain temporal measurements on time scales from seconds or less to years or longer. Some photometric output products, such as differential photometry, require fewer additional steps, whereas to obtain the absolute flux for an object, additional CCD frames of photometric standards are needed. These standard star frames are used to correct for the Earth's atmosphere, color terms, and other possible sources of extinction that may be peculiar to a given observing site or a certain time of year (Pecker, 1970).

We start this chapter with a brief discussion of the basic methods of performing photometry when using digital data from 2-D arrays. It will be assumed here that the CCD images being operated on have already been reduced and calibrated as described in detail in the previous chapter. We will see that photometric measurements require that we accomplish only a few steps to provide output flux values. Additional steps are then required to produce light curves or absolute fluxes. Remember, for a photometrist, every photon counts but the trick is to count every photon.

As an introduction to the level of atmospheric extinction one might expect as a function of observational elevation and wavelength, Table 5.1 lists values of the extinction in magnitudes resulting from the Earth's atmosphere for an observing site at 2200 m elevation. Note that for observations made at reasonable airmass and redward of 4000 Å, the effect of the Earth's atmosphere is, at worst, a few tenths of a magnitude. The details of photometric corrections

Table 5.1. *Example Atmospheric Extinction Values (Magnitudes)*

Altitude	Airmass	3000 Å	3500 Å	4000 Å	4500 Å	5000 Å	5500 Å	6000 Å	6500 Å	7000 Å	8000 Å	9000 Å	10 000 Å
90	1.00	1.2	0.65	0.4	0.3	0.2	0.2	0.2	0.1	0.1	0.1	0.1	0.1
75	1.04	1.2	0.65	0.4	0.3	0.2	0.2	0.2	0.1	0.1	0.1	0.1	0.1
60	1.15	1.3	0.75	0.5	0.3	0.3	0.2	0.2	0.2	0.1	0.1	0.1	0.1
45	1.41	1.6	0.9	0.6	0.4	0.3	0.3	0.2	0.2	0.2	0.1	0.1	0.1
30	1.99	2.3	1.3	0.8	0.6	0.4	0.4	0.3	0.3	0.2	0.2	0.1	0.1
20	2.90	3.3	1.55	1.2	0.8	0.6	0.5	0.5	0.4	0.3	0.2	0.2	0.2
15	3.82	4.4	2.5	1.6	1.1	0.8	0.7	0.6	0.5	0.4	0.3	0.3	0.2
10	5.60	6.4	3.65	2.3	1.6	1.2	1.1	1.0	0.7	0.6	0.4	0.4	0.3
5	10.21	11.8	6.7	4.2	2.9	2.2	1.9	1.7	1.4	1.1	0.8	0.7	0.6

for the Earth's atmosphere and extinction effects are not germane to the topic of this book and their discussion here would be beyond the allowed space limitations. The interested reader is referred to the excellent presentations in Young (1974), Hendon & Kaitchuck (1982), Dacosta (1992), and Romanishin (2004). Further good discussions of photometric data handling are presented in Hiltner (1962), Howell & Jacoby (1986), Stetson (1987), Walker (1990), Howell (1992), Merline & Howell (1995), and Howell, Everett, & Ousley (1999).

5.1 Stellar photometry from digital images

Prior to the time when CCDs became generally available to the astronomical community, digital images of astronomical objects were being produced by detectors such as silicon intensified targets (SITs), video tube-type cameras, image tubes, and electronographic cameras. In addition, scanning of photographic plates with a microdensitometer resulted in large amounts of digital output. These mechanisms produced digital data in quantity and at rates far in excess of the ability of workers to individually examine each object of interest within each image. Today, the amount of CCD data greatly exceeds this limit. Thus, in the early 1980s, work began in earnest to develop methods by which photometry could be obtained from digital images in a robust, mostly automated manner.

One of the first such software packages to deal with digital images was written by Adams, Christian, Mould, Stryker, and Tody (Adams *et al.*, 1980) in 1980. Numerous other papers and guides have been produced over the years containing methods, ideas, entire software packages that perform photometry, and specific detailed information for certain types of objects. I have tried to collect a fairly complete list of these in Appendix A. While details vary, the basic photometric toolbox must contain methods that perform at least the following primary tasks: (i) image centering, (ii) estimation of the background (sky) level, and (iii) calculation of the flux contained within the object of interest. We will assume below, for simplicity, that we are working with stellar images that to a good approximation are well represented by a point-spread function of more-or-less Gaussian shape. Deviations from this idealistic assumption and nonpoint source photometry will be discussed as they arise.

5.1.1 Image centering

Probably the simplest and most widely used centering approximation for a point-spread function (PSF) is that of marginal sums or first moment distributions. Starting with a rough pointer to the position of the center of the star (e.g., the cursor position, reading off the x, y coordinates, or even a good guess), the intensity values of each pixel within a small box centered on the image and of size $2L+1 \times 2L+1$ (where L is comparable to the size of the PSF) are summed in both x and y directions (see Figure 5.1). The x, y center is computed as follows: the marginal distributions of the PSF are found from

$$I_i = \sum_{j=-L}^{j=L} I_{i,j}$$

and

$$J_j = \sum_{i=-L}^{i=L} I_{i,j},$$

where $I_{i,j}$ is the intensity (in ADU) at each x, y pixel; the mean intensities are determined from

$$\bar{I} = \frac{1}{2L+1} \sum_{i=-L}^{i=L} I_i$$

and

$$\bar{J} = \frac{1}{2L+1} \sum_{j=-L}^{j=L} J_j;$$

and finally the intensity weighted centroid is determined using

$$x_c = \frac{\sum_{i=-L}^{i=L}(I_i - \bar{I})x_i}{\sum_{i=-L}^{i=L}(I_i - \bar{I})}$$

for all $I_i - \bar{I} > 0$ and

$$y_c = \frac{\sum_{j=-L}^{j=L}(J_j - \bar{J})y_j}{\sum_{j=-L}^{j=L}(J_j - \bar{J})}$$

for all $J_j - \bar{J} > 0$.

For well-sampled (see Section 5.9), relatively good S/N (see Section 4.4) images, simple x, y centroiding provides a very good determination of the center position of a PSF, possibly as good as one fifth of a pixel. More sophisticated schemes to provide better estimations of image centers or applications appropriate to various types of non-Gaussian PSFs are given in Chiu (1977), Penny & Dickens (1986), Stone (1989), Lasker *et al.* (1990b), Massey & Davis (1992), Davis (1994), and Howell *et al.* (1996).

5.1.2 Estimation of background

The importance of properly estimating the background level on a CCD resides in the fact that the same pixels that collect photons of interest from an astronomical source also collect photons from the "sky" or background, which

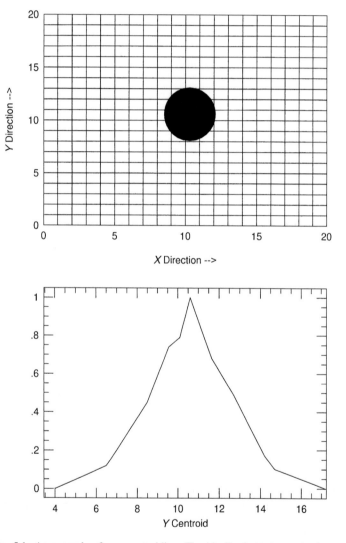

Fig. 5.1. An example of x, y centroiding. The idealized star image in the top box sits on a pixel grid with a center of $(x, y) = (10.3, 10.6)$. The two other plots represent the x, y centroids for the star image normalized to a maximum height of one. In this case, the star center is well approximated by the peaks in the x, y centroids.

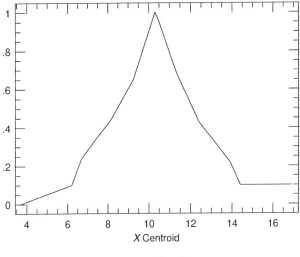

X Centroid

Fig. 5.1. (cont.)

are of no interest. Remember that the background or sky value in a CCD image contains not only actual photons from the sky but also photons from unresolved astronomical objects, read noise, thermally generated dark current electrons, and other sources. All of these unwanted additional photons must be accounted for in some manner, estimated, and removed from the image before a final determination of the source flux is made. In order to determine this background level, a common technique is to place a software annulus around the source of interest and then use statistical analysis to estimate its mean level on a per pixel basis.

The background or sky annulus is usually defined by an inner and outer radius or by an inner radius and a width (see Figure 5.2). One simple, yet powerful, manner by which an estimation of the background level can be made is simply to extract the values of all the pixels within the annulus, sum them, and divide by the total number of pixels within the annulus. This provides an average value per pixel for the background level of the CCD image. For a good statistical determination of the background level, the total number of pixels contained within this annulus should be relatively large, about three times the number contained within that of the source aperture (Merline & Howell, 1995).[1] A more robust estimator, requiring very little

---

[1] Partial pixels arising from placing a circular annulus on a rectangular grid are usually not of concern here, as the number of annulus pixels is large. However, partial pixels cannot be so easily dismissed when we are determining the intensity within the much smaller source aperture.

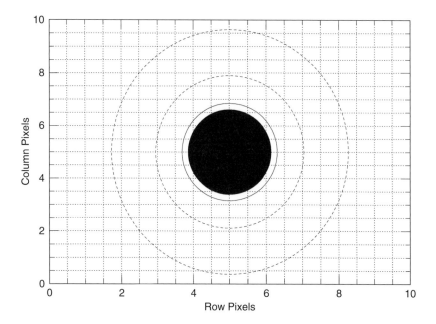

Fig. 5.2. Schematic drawing of a stellar image on a CCD pixel grid. The figure shows the location of the star, the "star" aperture (solid line), and the inner and outer "sky" annuli (dashed circles).

additional work, is to collect all the pixel values from inside the sky annulus, order them in increasing value, and find the median intensity, B_M.[1] A nice touch here is to reexamine the list of annulus pixel values and toss out all those with values greater than $\pm 3\sigma$ from B_M. This last step will eliminate cosmic ray hits, bad pixels, and contamination from close-by astronomical neighbors if they exist.

When applying a median filter and the 3σ cutoff technique to the list of background pixels, one can use the remaining annulus pixel values to construct a background histogram computed with a bin width resolution of say 0.5 ADU (Figure 5.3). The background histogram will be centered on the median background value with all contained pixel values within $\pm 3\sigma$. Since the detailed resolution of 0.5 ADU binning will likely produce a ragged histogram (since only a finite number of background pixels is used), some smoothing of the histogram may be useful, such as that done by Lucy (1975).

---

[1] Note that the statistical values determined from a CCD image for the median or the mode must pick their answer from the list of the actual CCD pixel ADU values, that is, values that are integers containing only digitized levels and thus digitization noise. The statistical mean, however, allows for noninteger answers. This seemingly subtle comment is of great importance when dealing with partial pixels, undersampled data, or high CCD gain values.

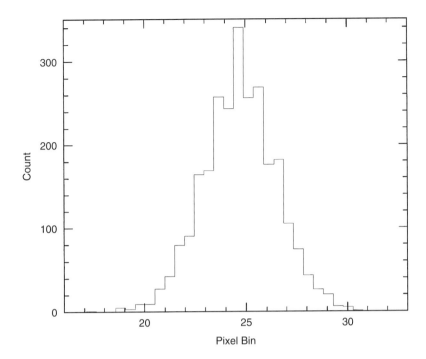

Fig. 5.3. Histogram of the "sky" annulus around a star in the CCD image shown in Figure 4.5. Notice the roughly Gaussian shape to the sky distribution but with an extended tail toward larger values. This tail is due to pixels that were not completely calibrated in the reduction process, pixels with possible contamination due to dark current or cosmic rays, pixels with increased counts due to unresolved PSF wings from nearby stars, and contamination of sky annulus pixels by faint unresolved background objects. The need for some histogram smoothing, such as that described in the text, is apparent, especially near the peak of the distribution.

Lucy smoothing will broaden the histogram distribution slightly after one iteration but application of a second iteration will restore the correct shape and provide a smooth histogram from which to proceed. This final step produces a statistically valid and robust estimator from which we can now compute the mean value of the background, $\bar{B}$. A determination of the centroid of the final smoothed histogram, using equational forms for the centroid in one dimension such as those discussed above, can now be applied. Centroiding of this smoothed histogram is not influenced by asymmetries that may have been present in the wings of the initially unsmoothed values.

   Correct estimation of the level of the CCD background on a per pixel basis is of increasing importance as the S/N of the data decreases and/or the CCD pixel sampling becomes poor. A background level estimation for each pixel that is off by as little as a few ADU can have large effects on the final result (Howell, 1989). Determination of the CCD background level has an associated error term $\propto (1 + n_{\mathrm{pix}}/n_B)^{-1/2}$, which should be included in the S/N calculation of the final result. The "sky" is the limit.

5.1.3 Estimation of point source intensity

We now come to the pixel values of interest, that is, those that contain photons from the source itself. Using a software aperture of radius r centered on the x, y position of the centroid of the source PSF, we can extract the values from all pixels within the area $A(= \pi r^2)$ and sum them to form the quantity S, the total integrated photometric source signal. The sum S contains contributions from the source but also from the underlying background sources within A. To remove the estimated contribution to S from the background, we can make use of the value $\bar{B}$, discussed above. We can calculate an estimate of the collected source intensity, I, as $I = S - n_{\mathrm{pix}} \bar{B}$, where n_{pix} is the total number of pixels contained within the area A. There are some additional minor considerations concerning this procedure but these will not be discussed here (Merline & Howell, 1995).

A final step usually performed on the quantity I, which we will discuss further below, is to determine a source magnitude. The value of a magnitude is defined by the following standard equation:

$$\text{Magnitude} = -2.5 \log_{10}(I) + C,$$

where I is the source intensity per unit time, that is, the flux (universally given as per second), and C is an appropriate constant (usually ~ 23.5–26 for most earthly observing sites) and determined in such a manner so that the calculated source magnitude is placed on a standard magnitude scale such as that of the Johnson system or the Strömgren system.

As we mentioned above, when using circular apertures on rectangular pixel grids, partial pixels are inevitable. While we could toss them away for the large background area, we cannot follow a similar sloppy procedure for the smaller source aperture. Thus the question becomes, how do we handle partial pixels? This is not a simple question to answer and each photometric software package has its own methodology and approach. The three choices a software package can use are:

1. Do not use partial pixels at all. Any source intensity that falls into the source aperture but within a partially inscribed pixel is simply not used in the calculation of S.

2. Sum the values for every pixel within the source aperture regardless of how much or how little of the pixel area actually lies within *A*.
3. Make use of some computational weighting scheme that decides, in a predefined manner, how to deal with the counts contained within each partial pixel in the source aperture.

This last choice often uses the ratio of the pixel area inside the source aperture to that outside the aperture as a simple weighting factor. A computational scheme to handle partial pixels in a software package designed to perform digital photometry is the hardest of the above choices to implement, but it will provide the best overall final results. To know exactly how a certain software package handles partial pixels, the user is referred to the details presented within the documentation provided with the software. Many PC-type packages that perform photometry on CCD images do not detail their partial pixel and magnitude calculation methods and are therefore "black boxes" to be avoided.

There are two basic methods by which most observers estimate the total integrated signal within their source aperture: point-spread function fitting and digital aperture photometry. The first method relies on fitting a 2-D function to the observed PSF and using the integrated value underneath this fitted function as an estimate of *S*. The second method, digital aperture photometry, attempts to place a software aperture about the source profile (as shown in Figure 5.2), centered in some manner (e.g., *x*, *y* centroids), and then simply sums the pixel values within the source aperture to provide the estimation of *S*. We will discuss each of these methods in turn below and note here that it is unlikely that a single method of estimation will be the best to use in all possible situations. For example, for severely undersampled data the method of pixel mask fitting (Howell *et al.*, 1996) provides the best solution.

5.2 Two-dimensional profile fitting

The profiles of astronomical point sources that are imaged on two-dimensional arrays are commonly referred to as point-spread functions or PSFs. In order to perform measurements on such images, one method of attack is profile fitting. PSFs can be modeled by a number of mathematical functions, the most common include Gaussian,

$$G(r) \propto e^{\left(\frac{r^2}{2a^2}\right)},$$

modified Lorentzian,

$$L(r) \propto \frac{1}{1 + (r^2/a^2)^b},$$

and Moffat,

$$M(r) \propto \frac{1}{(1+r^2/a^2)^b}$$

representations, where r is the distance from the center of the point source and a and b are fitting parameters (Stetson, Davis, & Crabtree, 1990). These types of functional forms can be used to define the PSF for each star within an image by the assumption that they provide a good representation of the data themselves. For example, adjustment of the values of a and b within one of these functions may allow an imaged PSF to be matched well in radius and profile shape (height and width), allowing a simple integration to be performed to measure the underlying flux.

Generally, the above functions are only a fair match to actual PSFs and so a second method of profile fitting can be applied. This method consists of using an empirical PSF fit to the actual digital data themselves, producing modified versions of the above functions. Depending on the application, PSFs may be evaluated at the center of a pixel or integrated over the area of each pixel. Even more general methods of allowing the data to produce completely analytic forms for the PSF functions have been attempted. The techniques and use of empirical PSFs could fill an entire chapter; we refer the reader to King (1971), Diego (1985), and Stetson (1987) for more details.

Both techniques, the use of completely mathematical forms for a PSF approximation and the more empirical method, have their advantages and disadvantages. Model PSF fitting allows the necessary integrations and pixel interpolations to be carried out easily as the functions are well known, while the empirical method, which makes hardly any assumptions about the actual shape of the PSF, is only defined on the CCD pixel grid and not in any general mathematical way. This latter complication can cause difficulties when trying to interpolate the empirical shape of one PSF somewhere on the CCD (say a bright reference star) to a PSF located somewhere else on the same image but that is likely to have a different pixel registration. For this reason, some implementations of empirical PSF fitting actually make use of the sum of an analytic function (such as one of those given as above) and a look-up table of residuals between the actual PSF and the fitting function. Figure 5.4 shows examples of some PSF models and some actual PSFs obtained with CCDs.

Procedurally, profile fitting techniques work by matching the implied PSF to the actual digital data in a 2-D fashion and within some radius, r, called the fitting radius. An attempt is then made to maximize some goodness-of-fit criteria between the assumed PSF and the observed one. PSF fitting can be further optimized by fitting N point sources within the image simultaneously

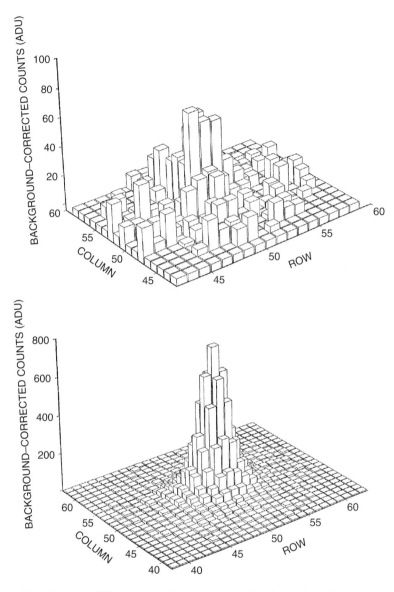

Fig. 5.4. Stellar PSFs are shown for various cases. The figure above shows two model PSFs, one for a bright star (S/N ~125) and one for a faint star (S/N ~ 20). The remaining two panels show similar brightness stars but are actual CCD data. Note that the models are shown as 3-D pixel histograms whereas the real data are represented as spline profile fits to the actual PSFs. The disadvantage of the latter type of plotting is that the true pixel nature of the image is lost.

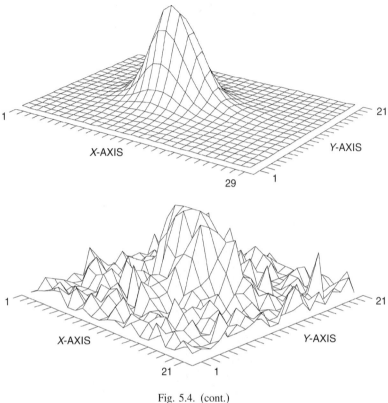

Fig. 5.4. (cont.)

(usually one uses the brightest stars within the image) and using some combination of statistically weighted mean values for the final fitting parameters.
PSF fitting can be very computationally demanding, much more so than the
method of aperture photometry discussed below. However, for some types of
imagery, for example crowded fields such as within star clusters for which
some PSFs may overlap, PSF fitting may be the only method capable of
producing scientifically valid results (Stetson, 1987, 1992, 1995).

5.3 Difference image photometry

One method used today for studies of photometric variability is the technique
of difference image photometry (DIA), also called difference image analysis
or image subtraction (Tomaney & Crotts, 1996). DIA is useful in (very)
crowded field photometry or when searching for variable sources that may
be blended with other possibly nonpoint sources. Most modern photometric
searches for supernovae in external galaxies or gravitational lenses use DIA

as these studies involve imaging in very crowded stellar fields and searching for highly blended sources.

The basic idea of DIA is to take a reference image and subtract from it images of the same field of view but taken at different times. An example would be to take a CCD image of a star cluster at an airmass of one and use this as your reference image. Additional images taken of this same field over time are then each subtracted from the reference image and variable sources show up in the difference image. Figure 5.5 shows a nice example of DIA from the supermacho project being carried out at the CTIO 4-m telescope in Chile. The reference image may actually be a sum of some number of the best images obtained (say during the best seeing) or an image observed at the lowest airmass. In practice, DIA is not so simple and involves setting up a CPU intensive, fairly complex software pipeline.

Before the simple process of subtraction from the reference image can occur, each successive image must be positionally registered, photometric normalized, and adjusted for other offending effects such as differential refraction, seeing and telescope focus changes, and possibly sky conditions. The matching of the point-spread functions between frames can be accomplished by Fourier divison (Alcock *et al.*, 1999) or a linear kernal decomposition in real space (Alard, 2000). Right away one can see that this is not a simple process. It involves setting up various transformation processes in software and a diligent eye to make sure they all work correctly in an automated

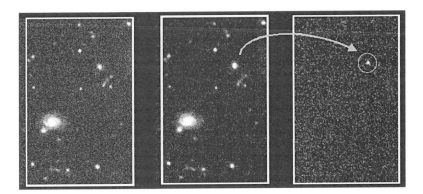

Fig. 5.5. An example of difference image analysis. The image was produced by C. Stubbs as a part of the high-z supernovae team using the supermacho data set. The image on the left is the reference image taken at epoch 1, the middle image is from epoch 2, 3 weeks later, and the right image is their difference. A supernova blended with its host galaxy image (middle frame) is clearly detected in the difference image.

fashion. Errors in these flux manipulation steps can be large and are often unaccounted for in the final result.

DIA is somewhat akin to profile fitting but is done on a frame-by-frame basis. The two-dimensional profile of each object in one frame is transformed to match those of the reference frame. One can model the profiles in a given frame using some small fraction of the brightest uncrowded sources per frame and applying the same model mapping of these sources to the reference frame to all objects in the given frame (Alcock *et al.*, 1999). This procedure saves CPU time but may introduce some uncertainty as it relies on a few of the brightest sources to be an exact match to the remaining sources. Spatial dependence, color dependence, pixel sampling, and seeing can all vary across an image and are hard to correct for perfectly. DIA has been used very effectively in a number of projects and continues to be the method of choice in certain regimes where crowding of some fashion is prevalent. The accuracy of the photometry delivered depends on how well the software processes of registration, convolution, and normalization are performed and what assumptions are used. For example, some DIA analysis assumes that stellar colors are well approximated by blackbody functions and in other cases that the bright stars well represent the remaining (fainter) stars. Both of these assumptions are valid to a point but are highly efficient in terms of processing the data. To date, photometric differences of 0.5 magnitude or better are easy to detect in a single difference frame. Thus DIA is a good technique for finding fairly large amplitude changes (i.e., newly brightened sources) but its ability to produce highly accurate light curves is yet to be fully explored.

The process of DIA is a mixture of photometry and astrometry plus profile fitting and using software to remap images to match the reference image. If observations are obtained of fields of interest that are not crowded or blended, than DIA is overkill and profile fitting or differential photometry (see below) work well and are easier to implement. These latter two techniques are also not subject to the addition of photon noise via the difference processing or any additional systematic effects as exist in DIA. But, as we note, every photometric technique has pros and cons and for highly blended or crowded fields DIA is a useful tool (see Tomaney & Crotts, 1996, Zebrun *et al.*, 2001, and references therein).

5.4 Aperture photometry

Aperture photometry is a technique that makes no assumption about the actual shape of the source PSF but simply collects and sums up the observed counts within a specified aperture centered on the source. The aperture used may be circular (usually the case for point sources), square, or any shape deemed useful. Aperture photometry is a simple technique, both computationally and conceptually, but this same simplicity may lead to errors if applied in an unsuitable manner or when profile fitting is more appropriate (e.g., severe blending).

The basic application of aperture photometry starts with an estimate of the center of the PSF and then inscribes a circular software aperture of radius r about that center. The radius r may simply be taken as three times the full-width at half-maximum ($r = 3 \cdot \mathrm{FWHM}$): the radius of a PSF that would contain 100% of the flux from the object (Figure 5.6) (Merline & Howell, 1995). Summing the counts collected by the CCD for all the pixels within the area $A = \pi r^2$, and removing the estimated background sky contribution within A, one finally arrives at an estimated value for I. We see again that partial pixels (a circular software aperture placed on a rectangular grid) are an issue, even for this simple technique. Using a square or rectangular aperture alleviates the need for involving partial pixels but may not provide the best estimate of the source flux. Noncircular apertures do not provide a good match to the 2-D areal footprint of a point source, thereby increasing the value of n_{pix} that must be used and decreasing the overall S/N of the measurement. Remember, however, that for bright sources, n_{pix} is essentially of no concern (see Section 4.4).

It has been shown (Howell, 1989; Howell, 1992) that there is a well-behaved relation between the radius of the aperture of extraction of a point source and the resultant S/N obtained for such a measurement. An optimum

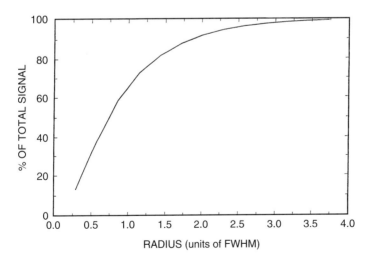

Fig. 5.6. For any reasonable PSF approximation, the figure above shows the run of the total encircled signal with radius of the PSF in FWHM units. Note that within a radius of $3 \cdot \mathrm{FWHM}$ essentially 100% of the signal is included.

radius aperture, that is, one that provides the optimum or best S/N for the
measurement, can be determined for any given PSF and generally has a
radius of near $1 \cdot$ FWHM. This optimum radius is a weak function of the
source brightness, becoming smaller in size for fainter sources. Figure 5.7
illustrates this idea for three point sources of different brightness.

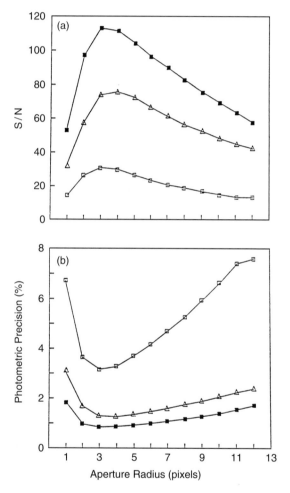

Fig. 5.7. The S/N obtained for the measurement of a point source is not constant
as a function of radius. There is an optimum radius at which the S/N will be a
maximum. The top panel shows this effect for three point sources that differ in
brightness by 0.3 (middle curve) and 2.0 (bottom curve) magnitudes compared
with the top curve (filled squares). The bottom panel presents the same three stars
as a function of their photometric precision. The image scale is 0.4 arcsec/pixel
and the seeing (FWHM) was near 1.2 arcsec. From Howell (1989).

To understand the idea of an optimum radius and why such a radius should exist, one simply has to examine the S/N equation given in Section 4.4 in some detail. To obtain a higher S/N for a given measurement, more signal needs to be collected. To collect more signal, one can use a larger aperture radius, up to the maximum of $3 \cdot$ FWHM. However, the larger r is, the more pixels that get included within the source aperture and the larger the value of n_{pix}. As n_{pix} increases, so does the contribution to the error term from noise sources other than the source itself. Thus, a balance between inclusion of more signal (larger r) and minimizing n_{pix} in the source aperture (smaller r) leads to an optimum extraction radius for a given source.

We saw in Figure 5.7 that if extracted at or very near an aperture radius of $3 \cdot$ FWHM, 100% of the light from a point source would be collected. However, to obtain the maximum S/N from your measurement, extraction at a smaller radius is warranted. If one extracts the source signal using an aperture that is smaller then the actual PSF radius itself, some of the source light that was collected by the CCD is not included in the summing process and is thus lost. This sounds like an incorrect methodology to use, but remember that inclusion of many pixels lying far from the source center also means inclusion of additional noise contributions to the aperture sum. Therefore, while one may wish to obtain the maximum S/N possible through the use of a smaller aperture (i.e., summation of less than the total collected source counts), for the final result it is often necessary to correct the answer obtained for this shortcoming.

In order to recover the "missing light," one can make use of the process of aperture corrections or growth curves as detailed by Howell (1989) and Stetson (1992). Growth curves do not make any demands on the underlying PSF except through the assumption that any bright stars used to define the aperture corrections are exact PSF replicas of any other (fainter) stars that are to be corrected. As we can see in Figure 5.8, growth curves for the brightest stars follow the same general shape, leading to minor or no necessary aperture corrections at a radius of $3 \cdot$ FHWM. The fainter stars, however, begin to deviate from the canonical growth curve at small radii, resulting in 0.5 up to 1.5 magnitudes of needed aperture correction. In general, as a point source becomes fainter, the wings of the source will contain pixels that have an increasingly larger error contribution from the background, leading to greater deviations from a master growth curve at large r, and thus a larger aperture correction will be needed.

As we will see below, if differential photometric results are desired, the aperture corrections used to bring the extracted source signal back to 100% are not necessary. This is only strictly true if all point sources of interest

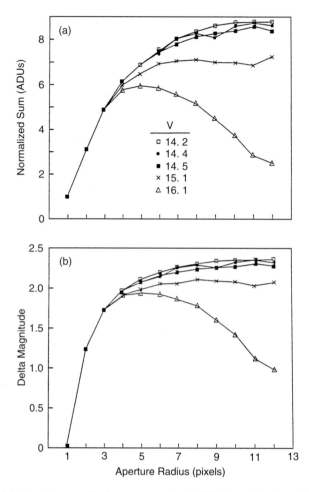

Fig. 5.8. Growth curves for five stars on a single CCD frame. The three brightest stars follow the same curve, which is very similar to the theoretical expectation as shown in Figure 5.6. The two faint stars start out in a similar manner, but eventually the background level is sufficient to overtake their PSF in the wings and they deviate strongly from the other three. Corrections, based on the bright stars, can be applied to these curves to obtain good estimates of their true brightnesses. The top panel presents growth curves as a function of normalized aperture sums while the bottom panel shows the curves as a function of magnitude differences within each successive aperture. The relative magnitudes of the point sources are given in the top panel and the image scale is the same as in Figure 5.6. From Howell (1989).

(those to be used in the differential measures) are extracted with the same (optimum) aperture and have identical PSFs. It is likely that on a given CCD image all stars of interest will not be of exactly the same brightness and will therefore not all have exactly the same optimum aperture radius (see Figure 5.7). Thus, a compromise is usually needed in which the extraction radius used for all sources of interest is set to that of the optimum size for the faintest stars. This procedure allows the faintest sources to produce their best possible S/N result while decreasing the S/N for bright stars only slightly. Another method is to use two or three different apertures (each best for 1/2 or 1/3 of the magnitude range) with the final differential light curves separated in halves or thirds by aperture radius (i.e., magnitude).

Advances in the technique of differential photometry have led to an output precision of 1 milli-magnitude for the brightest stars almost routinely. Everett and Howell (2001) outline the procedure in detail, providing the technique and equations to use and discuss a few "tricks" that help not only achieve very high precision but provide good results even for the faintest stars. The use of local ensembles of stars and production of an ensemble for every frame (not an average frame) are the main ones. Ensemble differential photometry is the method that provides the highest precision photometry one can obtain. This method will be used for the NASA Kepler Discovery mission to search for terrestrial size extra-solar planets, the GAIA Mission, and numerous ground-based time-resolved photometric surveys even in fairly crowded fields (e.g., Howell *et al.*, 2005; Tonry *et al.*, 2005)

5.5 Absolute versus differential photometry

Whether an observer needs to obtain absolute or differential photometric measurements depends on the objectives of the scientific and observational program. Absolute photometry results in a measurement of a given astronomical source leading to the true level of the flux received, say in ergs s^{-1}, or the total luminosity of a source in ergs, each within the specific band-pass used (e.g., a Johnson R filter). Differential photometry is just that: the final result is only known with respect to one or more other sources. This is a relative measurement, which, if the other source(s) are known in an absolute manner, can be placed on an absolute scale. All scientific measurements are really differential ones. The difference between absolute and differential measurements is simply that differencing from a known (in an absolute sense) source allows for an absolute result, whereas differencing from an unknown (in an absolute sense) source can only produce a relative result.

In photometric work one can view this in the following way. If the star Vega has its absolute flux known as a function of color (Tüg, White, & Lockwood, 1977), then a comparison of an unknown star with Vega would allow one to calibrate the unknown in an absolute sense. This differential measurement would produce an absolute result. However, if two unknown sources are compared over time, one may be seen to vary periodically allowing a scientific result to be produced, but its absolute flux is unknown. This is a differential measurement that produces only a relative result. If you are concerned about the true brightness of an object, say within a standard photometric system, or its color (e.g., B) in order to place it on an H–R diagram or to obtain its age or metallicity, then absolute photometry is needed. However, if you are after relative variability, say for periodicity analysis, or color indices (e.g., B–V) of a large group of stars such as a cluster in which you are looking for outliers, then differential photometry is likely to suffice.

A discussion of the observational and reduction techniques involved in absolute photometry is beyond the scope of this section. The reader is referred to any of the numerous books and articles that discuss the conversion of digital photometric data to (absolute) magnitudes and standard photometric systems (e.g., Hiltner (1962), Young (1974), Hendon & Kaitchuck (1982), Dacosta (1992)). The observational methodologies and equational transformations used to convert from CCD instrumental magnitudes to standard photometric system magnitudes are essentially identical to those used for similar observational programs with photomultiplier tubes (PMT). Absolute photometry from CCD data has been shown to be just as reliable and every bit as good as PMT measurements (Young, 1974; Walker, 1990; Kreidl, 1993). At the present, typical very good values for errors in the source magnitude for an *absolute* photometric result from CCD observations are ±1% or less.

Differential photometry concerns itself with the measurement of the difference in brightness of one or more astronomical sources when compared with one or more reference sources. With a two-dimensional CCD array, unlike the situation when using a PMT, it is often the case that multiple sources are imaged simultaneously and can therefore be used as references by which to compare the source(s) of interest. The assumption that is made in this simple approach is that the reference source(s) is not variable (at least over the time period of the observation) and the object(s) of interest can be compared with the reference source(s) on a frame-by-frame basis. The last step is important as it cancels out any seeing or atmospheric effects that may change with time.

The use of an ensemble of (bright) stars is common today when one is performing differential photometry. The brightest 20–50 stars in a CCD image (or a subregion of the CCD image) are averaged together, outliers removed, and the averaging process performed again to convergence. Proper statistical use of each star in the ensemble as well as weighting the error contribution each makes to the ensemble is required. The ensemble value in each frame is then used as the reference to which all other sources are compared. This process continues on every frame in the time series and at the end light curves for every source are produced (e.g., Howell *et al.*, 2005). No "calibration" for an offset between CCD images is needed as the ensemble (of the same stars) in each frame sets the base level and by definition this level will be equal across frames.

To date, the effects of differential refraction, color terms, and seeing changes do not seem to pose an issue for properly implemented ensemble differential photometry. In fact, the level of precision being reached is near the limit of linearity of A/D converters and often reveals small changes in the CCD gain caused by thermal effects. Higher precision photometry would benefit from more stable CCD electronics and controllers in order to reach its full extent. Until then, be aware that at ultra-high precision, additional calibration work may be needed to correct the output magnitudes at the sub milli-magnitude level.

Proper statistical comparison of the object(s) of interest and the reference(s) must take into account all sources of error as well as photon statistics in order to use differential photometry correctly (Howell & Jacoby, 1986; Howell, Mitchell, & Warnock, 1988; Honeycutt, 1992; Howell, 1992; Everett & Howell, 2001). Differential techniques allow one to obtain incredible precisions from CCD data. For example, photometric errors of ± 0.001 magnitude are easily obtainable today with differential techniques (Gilliland *et al.*, 1993; Howell, 1993; Howell *et al.*, 1999; Howell, Everett, & Ousley, 1999; Everett *et al.*, 2002; and Howell *et al.*, 2005).

5.6 High speed photometry

Many years ago (about 30) in the dark ages of astronomical instrumentation, photomultiplier tubes (PMTs) were making their farewell to astronomy and with them went almost all of the studies of high-speed phemomena. The Universe is a fast-paced place and many sources change on time scales of less than 1 second. Accretion phenomena and rotation of neutron stars are a few examples. Studies of such events and other fast changes are being resurrected thanks both to a renewed interest and to CCDs and instruments that are capable of the task.

Today, there are only a few modern instruments that can observe the Universe in under 1 second. They come in two varieties; non-CCD instruments, which use avalanche photodiodes or PMTs, and 2-D digital cameras, which use frame transfer CCDs (such as UltraCam) or shutterless readout of OTC-CDs (such as in OPTIC). UltraCam (Dhillon & Marsh, 2001) can simultaneously obtain three-color imaging photometry with integration times (actually frame transfer times, which keep a reasonable noise level) of 1 second. Subsecond readout (down to a few ms) is possible for smaller regions of interest. OPTIC (Howell *et al.*, 2003) can perform shutterless readouts as short as 1 ms for up to four regions of interest simultaneously. New versions of OTCCDs being produced for the WIYN observatory will increase this capability to include essentially unlimited regions across the entire field of view. Some other examples of high-speed photometric applications are presented in Schwope *et al.* (2004) and Nather & Mukadam (2004).

High speed photometry is often used to study fast phenomena such as stellar eclipses or short period pulsations. It is also useful to obtain ultra-high

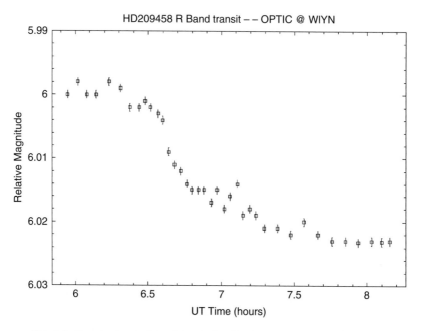

Fig. 5.9. Transit observation of the Jupiter-like extra-solar planet orbiting the V = 7.4 magnitude star HD209458. These R band photometric data were collected with the 3.5-m WIYN telescope using OPTIC and 1 ms integrations. Only the first half of the transit was measured as the observations were halted when the source reached an airmass of 2.5. Each CCD image collected over 10^6 photons and the time series has been co-added here to 58-s averages. The photometric precision per point is 0.0005 magnitudes. Data provided by the author.

precision photometry by allowing one to collect a large number of photons from (but not saturating) a bright source using rapid frame readout with later co-addition. Figure 5.9 illustrates this method as applied to the bright star HD209458, a G2V star with a Jupiter-like extra-solar planet. The planet transits the star every few days causing a $\sim 1.7\%$ drop in its light. While the transit event itself is easy to observe, even for small telescopes, much of the detailed astronomical information needed to compare with models of extra-solar planet atmospheres comes from the ingress and egress shape of the transit and from measurements of the transit depth to high precision.

5.7 PSF shaped photometry

The theoretically best PSF shape for good photometric results is a top hat or mesa shape. Point sources, however, come in only one shape, round, and they approximate a Gaussian distribution when imaged on a CCD. This is not the case anymore! OTCCDs have the ability to shift their collected charge in the CCD during integration in order to provide tip-tilt corrections (see Chapter 2). This same feature can be put to use to manipulate the incoming photons into whatever shape the user desires in order to increase the output science. Howell *et al.* (2003) used this property of OTCCDs to produce square stars (Figure 5.10).

The photometric precision available with square stars is generally greater than with normal stars as the PSFs are better sampled and contain higher S/N for a given source magnitude. The shaped stars have none of the drawbacks of similar techniques (i.e., defocusing of bright stars) but do render faint background sources and 2-D objects (such as galaxies) unusable (Tonry *et al.*, 2005). One of the great benefits of PSF shaped photometry is its ability to greatly increase the high photometric precision dynamic range of a CCD. Figure 5.11 shows the relationship between the recorded magnitude of a source and the photometric precision of its light curve obtained with ensemble differential techniques. The panel on the left is a conventional CCD (with normal star images) and the right hand side shows a similar result for a PSF shaped data set obtained with an OTCCD. The solid line in the left hand panel is the theoretically expected result using the CCD S/N equation (Chapter 4). The difference in the two plots is obvious revealing that the PSF shaped result yields the highest photometric precision over nearly 5 magnitudes of dynamic range (Howell *et al.*, 2005).

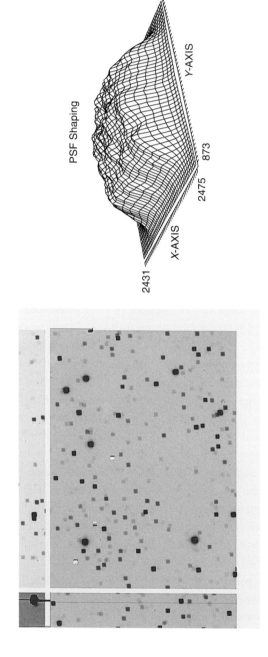

Fig. 5.10. PSF shaped star profiles using an OTCCD. The left hand panel shows a sub-section of an OTCCD image in which the stars were moved in a square shaped pattern (of 20×20 pixels) during each 300-second integration. The right hand panel shows a 3-D plot of one of the mesa-like PSFs.

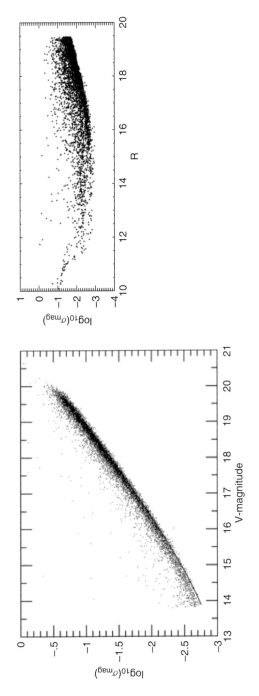

Fig. 5.11. Relationship between recorded stellar magnitude and the resulting photometric precision of the light curve for a sample of stars. The left panel is from a conventional CCD while the right panel is from an OTCCD using PSF shaping techniques. The shaped profiles (squares) provide similar high precision for about 5 magnitudes of brightness. Stars on the right, brighter than ~ 11, are saturated.

5.8 Astrometry

The science of the exact location and possible movement of astronomical sources has also been greatly advanced through the use of CCDs. Differential astrometric measurements of 1 milliarcsecond (mas) or better are achievable today, yielding new information on positions, proper motions, and parallaxes of astronomical objects. In differential astrometry, one measures the position of an object with respect to other objects within the same CCD field of view. This type of measurement is applicable to point source PSFs (e.g., stars) as well as moving objects (e.g., comets) and fuzzy objects (e.g., galaxies); however, the last two types of object have additional complexities not discussed here (Lasker *et al.*, 1990; Monet *et al.*, 1991; Monet, 1992).

Astrometric information is one method by which distance determination may be made for an object of interest. Of course the desire for a distance does not always translate into the ability to obtain such information. In general, in one night of obtaining a few good CCD observations, the determined position for a "well-behaved" star can be known to an astrometric accuracy of about 3 mas. A few tens of CCD frames taken over the course of a year (at a minimum of three different epochs) will allow astrometric accuracies near ±1 mas to be reached, while the best astrometric positions to date have errors of ±0.5 mas or less.

Astrometry performed with front-side illuminated CCDs will suffer from the effects of the overlying gate structures in ways such as we have discussed previously; that is, locating the exact center of the PSF is not without bias. The front-side gates must be traversed by incoming photons leading to a decreased (or no) blue response, intra-pixel variations in photometric (QE) response (Jorden, Deltorn, & Oates, 1994), and the need to use color terms to adjust the measured PSF centroid position to that of the actual source position. We will discuss an interesting CCD effect caused by intra-pixel QE variations in the next chapter.

Back-side illuminated CCDs present a different situation for astrometric work entirely. As mentioned earlier, their response to blue light is much improved and no major source of intra-pixel deviation exists. The physical flatness of a back-side illuminated CCD over its surface can be of concern and can introduce an additional term to be corrected for when measuring source positions and relative source offsets. Additionally, thinned CCDs may allow very long wavelength light to pass entirely through the device, be reflected back into the CCD itself, and be collected at a slightly different location (e.g., a neighboring pixel) from its original incoming path location. This long wavelength reflection effect can cause a slight blurring of the measured light from the source, a particularly worrisome issue for astrometry of very late type stars. The best astrometric measurements to date, that is, the ones with the smallest rms error and greatest repeatability, have all been made with thinned, back-side illuminated CCDs.

As with any CCD imaging or photometric observation, the use of a particular type of CCD and filter combination will produce a different effective wavelength of the imaged scene and thus a change in the relative positions of different sources as a function of the zenith distance of the observation and the color (spectral type) of each individual source. There is no simple or even complete solution to this issue of differential color refraction but there are some items to note.[1] The use of narrow-band filters in front of your CCD is of help in this situation because such filters greatly restrict the band-pass of the observation, thereby reducing the effects of differential refraction and color terms. However, the use of narrow-band filters is usually impractical due to the large loss of incoming signal from the astronomical sources of interest. Astrometric observations made in long wavelength (red) band-passes have merit as they eliminate many of the refractive effects to start with. Finally, CCD observations obtained near source meridian passage are also a plus given that for an airmass of 1.0, refractive changes are essentially zero within the entire CCD field of view. Details of astrometric observations and a discussion of such effects is presented in Monet & Dahn (1983), Monet *et al.* (1991), and Girard *et al.* (2004).

Issues of related concern for precision astrometry involve image focus, seeing, use of different filters, PSF stability, telescope collimation, and many others (Monet, 1992). Detailed information concerning one's CCD is vital when attempting precision astrometry. For example, the standard pixel size value available in most of the literature for a 15-micron pixel TI CCD states that the pixel size is 15 microns across. However, the true pixel size is 15.24 microns, a difference of one quarter of a pixel, a value that can cause significant errors in precise astrometric measurements. Finally, even more subtle effects such as nonuniformly spaced pixels and improperly produced pixels that differ slightly ($\sim$1%) in size must be considered. A recent astrometric solution for the HST WFPC2 CCDs produces an accuracy to $\pm$0.1–0.2 pixels and appears to be limited here due to telescope breathing, variations across the filters, and even physical movement of the CCDs themselves (Anderson & King, 2003). The bottom line is, as in all cases where highly precise results are desired, one must know thy CCD in great detail.

---

[1] While reading the remainder of this paragraph, the reader may wish to skip ahead and glance at Table 6.2.

Data reduction methods of CCD images from which astrometric measures are to be obtained are similar to those discussed above for general CCD imagery. The differences in the process occur when the final reduced frames are to be used to produce output astrometric information. The interested reader is referred to the discussions and results given in Monet & Dahn (1983), Lasker *et al.* (1990), Monet *et al.* (1991), Monet (1992), and Girard *et al.* (2004).

5.9 Pixel sampling

An important consideration in photometric and astrometric measurements made with a CCD is how well the PSF is sampled on the two-dimensional array. PSFs that are well sampled by a CCD observation will lead directly to the result that the center and shape of the PSF will be known to higher precision, and thus one will obtain a final answer that will be of higher accuracy. We can define a sampling parameter, r, as follows (Howell *et al.*, 1996; Buonanno & Iannicola, 1989):

$$r = \frac{\text{FWHM}}{p},$$

where FWHM is the full-width half-maximum value of the source PSF and p is the pixel size, both values given in the same units. For r less than about 1.5, digital data are considered undersampled. As can be seen from the above expression, r will be small for the case of a CCD with large pixel sizes compared with the total areal coverage of the imaged PSF. The other possible case of small r values is if the CCD image contains very tight PSFs such as those that might be obtained at observing sites with very good seeing, if using adaptive optics systems, or for CCD images obtained outside the Earth's atmosphere (i.e., space-based telescopes). Real life examples of cases that will produce undersampled images (i.e., small r values) are a typical wide-field telescope outfitted with a large-format CCD, such as a Schmidt telescope or a camera lens, or a space-based telescope such as the Hubble Space Telescope wide-field planetary camera (WFPC) (Holtzman, 1990; Howell *et al.*, 1996, and Section 7.1).

Anytime the value of r approaches the limiting case of undersampling, standard software methods and techniques of astrometric and photometric data analysis will begin to produce increasingly larger errors and poorer fits as r decreases further (see Section 7.1). Photometric and astrometric errors obtained from CCD observations are related in that the analysis techniques for each type of measurement are very similar. Both photometry and astrometry require intimate knowledge of the centroid position of the source PSF. However, it has been shown (King, 1983; Stone, 1989; Howell & Merline, 1991) that for undersampled data the photometric error is least for source PSFs

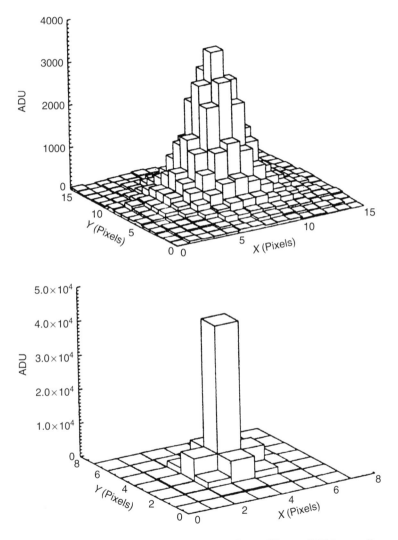

Fig. 5.12. The effects of pixel sampling are shown. The top PSF is a well-sampled star image with a S/N of near 230. The bottom panel shows the same PSF but now severely undersampled and centered at the middle of a pixel and (next page) at the corner of four pixels respectively. Note that the undersampled profiles are not well represented by a Gaussian function. From Howell *et al.* (1996).

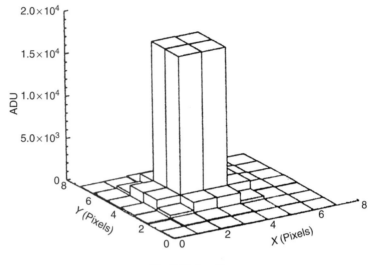

Fig. 5.12 (cont.)

that are centered on the edge of a pixel or exactly at its center, whereas for astrometric data, the resulting error is least when the source PSF is centered midway between a pixel's edge and its center.

The rule of thumb for pixel sampling on a CCD follows directly from the statistical result of Nyquist sampling. That is, sampling of the PSF of an astronomical source will be optimal in terms of S/N, error rejection, data analysis, and so on for a source PSF that has its FWHM value sampled over about two pixels (i.e., FWHM $\sim$2 · pixel size). For example, if the average seeing at an observing site produces source PSFs with FWHM values of 2 arcsec, then an ideal (optimal) CCD pixel size to use would be one for which each pixel within the array has a projected image size of 1 arcsec across.[1] A rigorous mathematical definition of undersampling, based on the Nyquist theorem, identifies critical sampling as the sampling interval that is equal to the width (i.e., standard deviation) of the PSF. For a Gaussian PSF this corresponds to a FHWM equal to 2.355 pixels. Of course an ideal image scale is hard to meet in reality as seeing and telescope focus change with time, source PSFs generally do not fall onto the CCD pixel grid exactly on a pixel boundary, and one generally has only a limited number of available CCD cameras with fixed pixel sizes.

---

[1] The determination and measurement of CCD pixel size or plate scale was discussed in Section 4.1.

Since CCD detectors do indeed sample astronomical sources in a quantized manner, pixel sampling within the array will cause even identical PSFs to change their detailed appearance slightly, even within the same observation. The effects of such sampling differences become worse as the sampling parameter (r) becomes smaller (Howell & Merline, 1991; Merline & Howell, 1995). Figure 5.12 illustrates some examples of various values of r caused by CCD pixel size. The top panel shows a well-sampled source PSF that appears to be more or less Gaussian in shape. The remaining two panels in Figure 5.12 show the same model PSF but now imaged as a poorly sampled ($r = 1$) source. The undersampled cases are for a source PSF with a pixel-centered centroid and a corner-centered centroid respectively. Notice in Figure 5.12 that the undersampled PSFs are not well represented by a Gaussian function (Buonanno & Iannicola, 1989; Holtzman, 1990; Howell & Merline, 1991; Howell *et al.*, 1996).

5.10 Exercises

1. How do the numbers in Table 5.1 help us understand the fact that the sun appears red at sunrise and sunset?
2. At what wavelength would you choose to observe an astronomical source if you could only point your telescope at an elevation of 20 degrees?
3. Describe in words what a point-spread function is. Do extended objects, such as galaxies, have PSF's? If they do, what do they look like?
4. Write a computer program to find the center of an astronomical image that has a circular shape (and radial light profile) when projected on a 2-D surface. Try various techniques and determine: which centering method is the fastest? Which is the best? What if the shape were triangular? A square?
5. Write a computer program to calculate the PSF of a star imaged on a CCD. Use the information in the previous chapter to include all noises in the calculation.
6. Using Figure 5.2, describe the need for a star aperture, a background annulus, and why they are circular. Would other shapes work as well?
7. Perform a qualitative analysis of why, or why not, partial pixels are important in stellar photometry. How does your answer relate to the CCD pixel size and plate scale?
8. Using the results presented in Figure 5.3, discuss quantitatively the error term $(1 + n_{pix}/n_B)^{1/2}$. This term was introduced in Chapter 4.
9. Read the paper by Tüg *et al.* (1977). What is so important about the star Vega? Can you design a similar experiment to try with your CCD camera?

10. Make plots of each of the model PSFs (given as equational representations in Section 5.2). How do they compare to the PSFs shown in Figure 5.4 and those in Diego (1985) and King (1971)? How do the pixel scale, plate scale, and plot type influence your answer?

11. Why is the FWHM of a PSF an important (dimensionless) value? What is so special about the value $3 \cdot$ FWHM? Explore the parameters "encircled energy" and "Strehl ratio" and relate them to the FWHM of a PSF. Which is a better descriptor of a PSF?

12. Discuss why PSF fitting photometry is the best method to use for the determination of absolute photometry of a source.

13. Make a table of pros and cons for PSF fitting, DIA, and aperture photometry. Which of these are differential measurements? Which are absolute? Discuss which is best to use for the following observational projects: Stellar photometry of variables in the LMC, supernovae searches in external galaxy clusters, light curve observations of a 20th magnitude star, and making a color magnitude diagram for an open cluster.

14. Determine the faintest source for which one can achieve a S/N of 50 in a photometric measurement lasting 1 second. Assume the image was obtained with a typical LBL CCD in the R band and used a 3.5-m telescope. Can you determine a simple relationship between telescope aperture, source brightness, and integration time for a given S/N?

15. Prove that a square star (top hat) profile provides the best S/N for a point source. What type of PSF shaping might be useful for extended object photometry?

16. At the current best limit of CCD astrometry, what is the furthest distance at which a star can be and still have its parallax measured?

17. Design an astrometric observational program that would provide ideal results for a sample of very blue objects. Discuss the type of CCD, telescope, and filters you would use. What if the program sample were red sources?

18. Discuss how different pixel sampling (over-sampled and under-sampled) would affect photometric observations with a CCD. Be specific for the methods of PSF fitting, DIA, and aperture photometry. How do these same parameters affect spectroscopic observations with a CCD? (Hint: if you get stuck here, read ahead to Chapter 6.)

6

Spectroscopy with CCDs

Although imaging and photometry have been and continue to be mainstays of astronomical observations, spectroscopy is indeed the premier method by which we can learn the physics that occurs within or near the object under study. Photographic plates obtained the first astronomical spectra of bright stars in the late nineteenth century, while the early twentieth century saw the hand-in-hand development of astronomical spectroscopy and atomic physics. Astronomical spectroscopy with photographic plates, or with some method of image enhancement placed in front of a photographic plate, has led to numerous discoveries and formed the basis for modern astrophysics. Astronomical spectra have also had a profound influence on the development of the fields of quantum mechanics and the physics of extreme environments.[1] The low quantum efficiency and nonlinear response of photographic plates placed the ultimate limiting factors on their use.

During the 1970s and early 1980s, astronomy saw the introduction of numerous electronic imaging devices, most of which were applied as detectors for spectroscopic observations. Television- type devices, diode arrays, and various silicon arrays such as Reticons were called into use. They were a step up from plates in a number of respects, one of which was their ability to image not only a spectrum of an object of interest, but, simultaneously, the nearby sky background spectrum as well – a feat not always possible with photographic plates. Additionally, and just as important, were the advantages of higher obtainable signal-to- noise ratios, higher quantum efficiency, and very good linearity over a large dynamic range. These advances permitted spectral observations of much fainter sources than were previously available. Two-dimensional spectroscopy allowed the large error contribution

---

[1] Extreme physical environments, such as very high temperatures, magnetic fields, and gravitational fields, are not often available within earthly laboratories.

from the often unknown sky background to be removed during data reduction procedures. However, the above electronic devices still had their problems. Use of high voltage, which caused distortions in the image, and the often low dynamic range both limited the ability to numerically resolve strong spectral lines from a weak continuum or to resolve weak lines in general.

The introduction of CCD detectors for use in astronomical spectroscopy was quick to follow their introduction into the field of astronomy itself. One of the first devices put into general use for astronomical spectroscopy was a Fairchild CCD placed into service at the Coudé feed telescope on Kitt Peak circa 1982. This author remembers that particular device and the remarkable advance it made to astronomy at the time. Observing without photographic plates was amazing.[1] The introduction of CCDs allowed test observations to be made, no chemical development was needed, you could view your data almost immediately after taking it, and mistakes caused you very little in lost time. In addition, fainter sources and better spectral resolution were easily obtained and caused a renaissance in astronomical spectroscopy. The fact that the Fairchild CCD had a read noise of 350 electrons seemed unimportant at the time.

We will begin our discussion of astronomical spectroscopy with point source observations. The term "point source" is generally taken to mean a star, but under various conditions, other objects can be observed as a point source. For example, an active galaxy does indeed often show extended structure in terms of spiral arms, but short exposures or observations intended to study only the nuclear regions are essentially point source measurements. A more formal definition might be that point sources are objects whose angular size is determined by the seeing disk or instrumental resolution. We will follow point source observations by introducing extended object spectroscopy. The major difference in these two types of spectroscopy is the type of output data product you end up with and the science obtained from the collected data. Our discussion here will concentrate more on the CCD aspects of astronomical spectroscopy with some discussion of the actual observational techniques and data reduction procedures. Various types of spectrographs and other related topics are discussed in detail in the excellent reviews given by Walker (1987), Pogge (1992), and Wagner (1992).

---

[1] For those readers interested in a bit of nostalgia, remember how one needed to cut the photographic plates completely in the dark, attempt to fit them into the plate holder licking one side along the way to find the emulsion, and then suffering the agony of defeat when you discovered that your 1 hour integration was made with the dark slide closed or the plate had been placed in the holder backwards!

6.1 Review of spectrographs

Spectroscopic observations can be thought of as a method by which one samples the emitted energy distribution from an astronomical source in wavelength bins of size $\Delta\lambda$. Broad-band filter photometry, for example, is a form of spectroscopy; it is merely one with extremely poor spectral resolution. To use spectral information to learn detailed physics for an astronomical object, one must be able to differentiate specific spectral features (lines) from the continuum within the observed spectrum and be able to make quantitative measurements of such features. Generally, this type of analysis requires a spectral resolution of at least 20–40 Å or better. Keep in mind, however, that various scientific objectives can be accomplished with varying amounts of spectral resolution. Schmidt telescope observations using an objective prism and imaging each spectrum onto a CCD have fairly low spectral resolution, but the imaged spectra are indeed useful if the purpose is to identify objects that have blue color excesses (see Section 6.7).

Figure 6.1 illustrates a typical astronomical spectrograph with the common components identified. An entrance slit, onto which the telescope focuses the

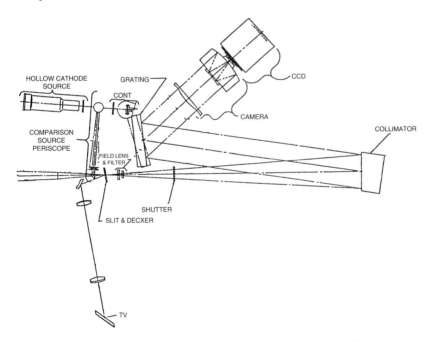

Fig. 6.1. Schematic diagram of a typical astronomical spectrograph. The major components are the CCD detector, the continuum and comparison line calibration sources, the TV slit viewer, and the grating. From Wagner (1992).

incoming light from the source of interest, is used both to set the spectral resolution and to eliminate unnecessary background light. An internal light source for the production of a flat field (called a projector flat in spectroscopy) and various wavelength calibration emission line sources are also included. These lamps usually consist of a quartz projector lamp for the flat fielding and a hollow cathode or arc lamp for the calibration sources. Both types of calibration lamp are included in the spectrograph in such a way as to attempt to make their light path through the slit and onto the CCD detector match as closely as possible that of the incoming telescope beam from an astronomical object. Some type of grating (commonly a concave reflection grating) is needed as the dispersive element, although a prism can also be used. Various camera optics, to re-image the slit onto the CCD detector and provide chromatic and field flatness corrections, finish the suite of standard components. Numerous variations on this standard theme have been and will continue to be used as cost, complexity, and purpose of the instrument are always issues.

Spectrographs with gratings (compared to prisms) and CCD detectors usually cover 1000–2000 Å of optical spectrum at a time with typical resolution of 0.1–10 Å/pixel. In order to cover more spectral range, a few observatories have built double spectrographs. These instruments consist of two separate spectrographs (each similar to that shown in Figure 6.1), which share the incoming light that is divided by a dichroic beam splitter into red and blue beams. Operating double spectrographs are discussed in Gillespie *et al.* (1995) and DePoy *et al.* (2004).

Two other astronomical spectrograph types are worth mentioning as they are increasingly used today. These are echelle type spectrographs and fiber fed spectrographs. Figure 6.2 shows an example of a cross-dispersed echelle spectrograph. This type of instrument provides high resolution spectroscopy (R = 50 000 to 100 000 or more) through the use of both an echelle grating to produce the high spectral resolution and a cross disperser to separate the orders and project them in two dimensions onto a CCD array. As an example of this type of observation, Figure 6.3 presents a cross-dispersed echelle spectrum of the Bpe star MWC162 obtained with the 6-m Bolshoi Teleskop Azimutalnyi (BTA) located in central Russia.

Astronomical spectrographs, of the types mentioned above, can also be fed by optical fibers that collect light at the telescope and bring it to the instrument. Many examples exist that use a single fiber to feed a table-mounted spectrograph, 10–30 fibers in a close bundle called an integral field unit (IFU), or cases of 100 or more fibers being placed in the focal plane at the telescope. The fibers collect light from individual objects at the focal

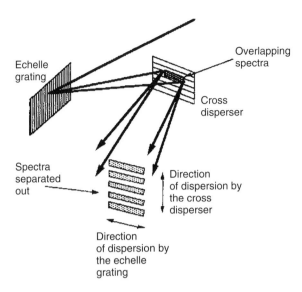

Fig. 6.2. Schematic diagram of a cross-dispersed echelle spectrograph showing the echelle grating and the cross disperser. The final 2-D spectral image is projected on to a CCD.

plane and carry the light as a fiber bundle to the awaiting spectrograph mounted on a table on the observatory floor or in an isolated room. Fiber fed spectrographs can provide ease for spectral observations (moving a single fiber into place is simpler than dismounting the current instrument and mounting a spectrograph), weight alleviation (a fiber is lighter than an instrument), or the ability to obtain multiple spectra at once (each fiber is positioned in the focal plane to observe one source). Figure 6.4 shows an example 2-D CCD image of nearly 70 spectra obtained simultaneously using the HYDRA multi-fiber spectrograph on the WIYN 3.5-m telescope at Kitt Peak National Observatory.

Let us define a few useful quantities in CCD spectroscopy. Table 6.1 lists the various definitions needed for use when discussing the optical properties of the telescope, collimator mirror (or lens), and the camera itself. Using the definitions in Table 6.1, we can define the magnification of the spectrograph as $M = F_{cam}/F_{col}$, the projected width of the slit at the CCD will be wM, and the slit width will subtend an angle on the sky of $\Theta = w/F$. The projected width of the slit at the CCD detector is then given by

$$r = wM = \Theta FM = \Theta fD\frac{F_{cam}}{F_{col}} = \frac{\Theta fDF_{cam}}{f_{col}d_{col}}.$$

Fig. 6.3. A cross-dispersed echelle CCD image of the Bpe star MWC162. The spectrogram covers 3900 Å to 5700 Å at high (echelle) dispersion with each order separated vertically by the cross disperser. Notice the presence of both emission and absorption lines as well as P Cygni profiles in the stronger Balmer lines. This image was obtained using the LYNX instrument on the BTA.

To avoid loss of efficiency within the spectrograph, the collimator focal ratio should match that of the telescope. To make this clear, we can write the projected slit width at the CCD detector as $\Re = \Theta D f_{\text{cam}}$. We assume here that the dispersing element does not change the collimated beam size.[1] If the slit is opened wide enough to allow all the light from a point source to pass through, the projected image size of the point source at the CCD detector is simply $\Re$.

The ability to separate closely spaced spectral features is determined by the resolution of the spectrograph. Spectral resolution is defined as $R = \lambda/\Delta\lambda$, in which $\Delta\lambda$ is the difference in wavelength between two closely spaced spectral features, say two spectral lines of equal intensity, each with approximate wavelength λ. Optical light spectral resolutions of a few hundred thousand to

---

[1] While not always true for diffraction gratings, this condition is realized for prisms.

Fig. 6.4. A multi-fiber spectrograph image obtained with HYDRA of a cluster
of galaxies. Each of the 70+ fibers was positioned at the focal plane to collect
the light of a single cluster member. The fiber bundle was then formed into a
linear array and each fiber's light passed through the spectrograph in a normal
fashion. The formed 2-D image of each spectrum was collected by a CCD and
presented to the user. Each spectrum will be extracted and examined separately.
Note how the spectral lines are nearly coincident for each object (as they are at
the same redshift and similar in type) but a few interlopers are present as well.

one million have been obtained for the Sun, whereas for typical astronomical
spectra, R is much less, being near a few thousand down to a few hundred
for very faint sources. For comparison, current R values in the infrared are
typically less than about 10 000.

Table 6.1. *Spectrograph definitions*

D	Diameter of Telescope
F	Focal Length of Telescope
f	Focal Ratio of Telescope
w	Width of Entrance Slit
d_{col}	Diameter of Collimating Mirror
F_{col}	Focal Length of Collimator
f_{col}	Focal Ratio of Collimator
d_{cam}	Diameter of Camera Mirror
F_{cam}	Focal Length of Camera
f_{cam}	Effective Focal Ratio of Camera

For further details of the actual components of astronomical spectrographs, various types of spectroscopy for different applications, and the way in which these components are used to produce spectra at various wavelengths and resolutions, see Robinson (1988b), DeVeny (1990), Pogge (1992), Wagner (1992), Cochran (1995), Corbally (1995), Quetoz (1995), and Stover *et al.* (1995).

6.2 CCD spectrographs

Current-day optical spectrographs almost exclusively use CCDs as their detector. The major reasons for this choice are the large free spectral range of modern CCDs (covering roughly 3000 to 11 000 Å), the linearity of the devices (better than 1% from zero counts to saturation) over a large dynamic range (allowing, for example, detection of absorption or emission lines as well as the continuum), and the large areal format of modern CCDs (2048 up to 4096 pixels or more in extent). This latter property is especially important for applications such as wide-field objective prism work with Schmidt Telescopes, multiobject fiber spectroscopy (Robinson, 1988c) in which the fiber fed spectra are placed in row order on the CCD, and for echelle spectroscopy for which many spectral orders are two-dimensionally imaged at once (Vogt & Penrod, 1988). The free spectral range obtained could, in principle, be as large as the detector's quantum response, but in practice limits in optical and grating design and CCD size restrict a single spectrograph coverage to somewhere near 4000 Å or less in the optical band-pass. As we have mentioned, some observatories have solved this limited spectral coverage issue by designing and using double spectrographs having two separate spectrograph arms (one red and one blue) with a CCD for each (Oke, 1988). A double spectrograph almost always uses a different type of CCD detector in each arm; each CCD is customized for the best possible detection properties for its particular wavelength coverage. Both high resolution ($R \sim 30\,000$–$80\,000$ or more) and low resolution ($R \sim$ a few thousand or less) spectroscopic applications are well suited to using CCDs as the detector; one simply has to match the CCD pixel size to the particular spectrograph and resolution being used.

Optimal sampling of a spectral line that is just unresolved occurs when the FWHM of the line is twice the physical pixel size (Nyquist sampling criteria). It can be assumed in this discussion that a spectral line has an approximately Gaussian shape for which a formal FWHM value can be determined. Note, however, that real spectral lines are not always this well behaved. In addition to matching the spectral line width to the pixel size, CCDs used for astronomical spectroscopy must also have very good charge transfer efficiency (CTE) in order to reduce smearing of spectral lines during readout, which would lead to a loss in spectral resolution. Also, small pixel sizes such as 15 or 9 microns are often desired to meet the Nyquist criteria discussed above.

Let us look at an example for a spectrograph that uses a CCD with 9-micron pixels as the detector. With this setup, the projected slit width size, $\mathfrak{R}$, must be near 18 microns to achieve optimal sampling. For an observing site with typical seeing of 1.5 arcsec, and using a 2- to 5-m telescope, we find (using the formulations given in Section 6.1) that the spectrograph camera must have a focal ratio near unity. This is a very fast focal ratio and requires excellent optical design and near perfect optical surfaces. For the CCD itself, this requirement means that its physical surface must be extremely flat throughout the entire extent of the chip (less than 0.5% rms for accurate spectrophotometry), in order to allow all parts of the spectrum to be in focus simultaneously. As we have seen, this level of flatness can be a difficult requirement to fulfill for certain types of CCD (e.g., thinned devices).

The above example for a CCD spectrograph informs us that, for large-aperture telescopes (say 8–12 m), optimum spectral sampling can only occur if some combination of the following conditions are met. As the telescope diameter increases, the camera focal length must decrease, the seeing disk must decrease, and the detector resolution element (2 CCD pixels) must increase in size. Currently, the hardest requirement to meet in this list is the design and construction of very fast focal ratio cameras. Increasing the CCD pixel size while retaining the large range of total wavelength coverage is a major driving force behind producing even larger format CCDs.

Exceptional seeing, less than 1 arcsec for example, would seem to be the dream of any spectroscopist. However, let us look at an example when very good seeing can cause unexpected results in CCD spectroscopy. The problem is as follows: CCDs are mounted in dewars and attached to the end of a spectrograph in some manner. The dewars are then aligned in an attempt to have the observed spectrum fall onto the detector along either its rows or columns. Perfect alignment of the CCD across the entire spectrum is rarely achieved and thus the imaged spectrum centroid must cross pixel boundaries in the dispersion direction (i.e., as a function of wavelength).

If the object seeing disk becomes less than the projected pixel size, the position of the spectral centroid falls within the pixel itself, alternately occurring at the center of some pixel and then at the pixel boundaries themselves. Wavelength-dependent QE effects within the pixels, due to their gate structures and intra-pixel "dead" spots, will cause apparent flux variations that can be as large as $\pm 10\%$ in amplitude. Additional complexities, such as which type of CCD is used, telescope focus, and tracking changes, are harder to quantify and correct for but can have similar effects. The problem described here, that of undersampling, can also occur in CCD imaging applications as well, when the majority of the source PSFs fall within a single pixel. Optimum sampling in CCD imaging also occurs at the Nyquist sampling limit, that is, a point source FWHM should be imaged across about two CCD pixels (see Section 5.9). A several percent error in CCD photometry can occur for images that are undersampled (Howell *et al.*, 1996; Holtzman, 1990).

Figure 6.5 shows an example of the ripple that can occur in a spectrum obtained under conditions of excellent seeing and for which intra-pixel QE effects are present (Rutten, Dhillon, & Horne, 1992). Possible methods of correction for spectral ripple include de-focus of the telescope or slightly trailing the spectrum up and down the slit during each integration,[1] neither of which are desirable. Corrections to the spectrum can also be applied after the fact using some empirical model during the reduction phase (Dhillon, Rutten, & Jorden, 1993; Jorden, Deltorn, & Oates, 1993). Spectral ripple, as well as our discussion of pixel sampling in the chapter on photometry, indicates that while very poor sampling is not ideal (as the collected image is spread out over many pixels, each of which adds unwanted noise) undersampling is not necessarily better. Optimum sampling is the best but is not always possible with a given instrument and CCD combination. In addition, the conditions of optimum sampling can change with time owing to effects such as seeing, telescope focus, the wavelength of light imaged, and other more subtle effects.

---

[1] This particular solution may remind us long-time observers of a Wobble Plate.

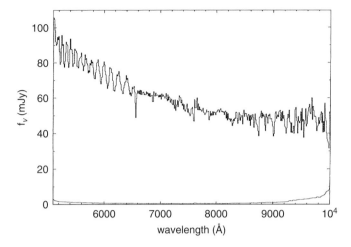

Fig. 6.5. Spectrum of the standard star Feige 92 taken during a time of excellent seeing. This 30-second exposure, obtained with the Faint Object Spectrograph on the Issac Newton telescope, has a pixel scale of 1.2 arcseconds/pixel and this spectrum was obtained when the seeing was sub-arcsecond. The ripple seen in the continuum near 6000 Å and blueward is due to intra-pixel QE differences, which are especially prominent for short wavelength photons in the EEV CCD used. From Dhillon, Rutten, & Jorden (1993).

The two-dimensional nature of a CCD allows one to place the spectrum anywhere on the array, targeting good regions of the CCD that avoid man-ufacturing flaws such as areas of poor sensitivity or dead pixels. Their 2-D design also provides the ability to simultaneously image the (nearby) sky spectrum with the object spectrum. Accurate sky subtraction from a source spectrum increases the S/N of the final result and allows much fainter sources to be observed. It also increases the reliability of the flux measurements and is probably the greatest factor making CCDs superior to other detectors as spectroscopic imagers.

Although large-format CCDs for spectroscopy (2048×2048, 1024×3072, etc.) are useful in many ways, they have some drawbacks as well. Large CCD formats take longer to readout (if windowing in not available), they make areal flat fielding critical (important for accurate fluxes and line profiles), and they provide more stringent restrictions on mounting, dewar design, and overall cost. Physically nonflat CCDs under vacuum (such as thinned devices), shifting of the CCD within the dewar, or movement of the larger, heavier dewar itself can all cause slight movement of spectral features, leading to errors in the analysis. Loss of LN2 from the dewar throughout the night has been identified as one cause of spectral movement (Smith, 1990a).

Readout times for large-format CCDs (typically 1–2 minutes, but faster in newer instruments) do not seem like such a big deal compared with typical integrations of 30–60 minutes or more. However, the longer the readout time, the lower the overall observational duty cycle one experiences, and, after obtaining all the necessary calibration spectra, flat fields, comparison arcs, and standard stars, this extra time can become costly. CCDs that have a limited dynamic range (those with small pixel sizes) force the user to make multiple (shorter) exposures for each type of needed image, especially calibration data for which the high S/N desired can easily saturate the detector. In addition, to produce the final spectrum one often wishes to co-add shorter object exposures to avoid numerous cosmic ray events that can hinder accurate flux and line profile measurements.

The finite readout time of large-format CCDs has led to numerous attempts to obtain high-speed spectroscopy via some technique that takes advantage of the two-dimensional nature of the detector. For example, one could step the detector, or the spectrum, along in a direction perpendicular to the dispersion at some predetermined rate, reading out the CCD only after an entire frame of individual spectra is collected. A better idea might be to use the fact that the collected charge can be moved back and forth electronically on the CCD from row to row, without actually reading it out. This movement is quite fast, say 200 rows in a few milliseconds, and suffers no readout noise penalty but only a slightly increased dark current. Spectra placed on the top 50 rows of a device will allow faster readout times or one could use the bottom 50 rows with electronic charge movement used to shift the spectrum upward between each integration time. Both of these processes have been tried with success (Robinson, 1988b; Skidmore *et al.*, 2004).

One could even slosh the charge back and forth periodically in order to buildup the signal in two spectra for say a given phase of a binary star. The spectrum not being exposed gets hidden under a mask and the final image produces two or more "phased" spectra. An idea such as this was developed for spectropolarimetry (Miller, Robinson, & Goodrich, 1988) with some success and led to the discovery of some previously unknown, yet interesting CCD effects (Janesick *et al.*, 1987a; Blouke *et al.*, 1988).

6.3 CCD spectroscopy

Two common measurements from an astronomical spectrum are those of the flux level as a function of wavelength and the shape and strength of spectral lines. Currently, the best determined absolute fluxes are near 1%, usually for spectral data covering a small wavelength range and for bright (high S/N) sources (e.g., Hayes & Latham (1975), Schectman & Hiltner (1976), Tüg, White, & Lockwood (1977), Oke & Gunn (1983), and Furenlid & Meylon (1990)). Absolute flux observations must make use of a wide slit or hole (10 arcsec or larger in size is typical) as the entrance aperture to the spectrograph. Use of such a wide slit assures that 100% of the source light is collected and avoids problems related to telescope tracking or guiding errors, seeing changes, and differential refraction (see below).

In principle, relative flux measurements should be determinable to even better levels than those quoted above and the shape of a spectral line should be almost completely set by the instrumental profile of the spectrograph. Spectroscopy desiring relative fluxes generally makes use of a narrow slit (matched to the seeing) to preserve the best possible spectral resolution. For example, for an observation requiring accurate radial velocity measurements a narrow slit would be preferred to achieve the highest possible spectral resolution. After all, the imaged spectral features (lines) are merely images of the slit itself focused onto the CCD.

Calibration of observed object spectra is normally performed by obtaining spectra of flux standard stars with the same instrumental setup including the slit width. During the reduction process, these standard stars are used to "flux" the object data, that is, assign relative fluxes to your object spectrum counts as a function of wavelength. It is assumed that during each integration, the object of interest and each standard star sent the same relative color and percentage of their PSF light per time interval through the slit and onto the CCD. Seeing changes, non-photometric conditions, guiding errors, and differential refraction can all negate this assumption.

Three factors unrelated to the CCD detector itself – tracking and guiding errors, seeing changes, and spectrograph slit angle – are important to your final spectral result in the following ways. Observations of an astronomical source made through a spectrograph slit are often obtained such that the slit size matches the object seeing disk. That is, the slit width is set to allow most of the PSF (for a point source) to pass through but kept small enough to eliminate as much of the sky background as possible from entering the spectrograph. Therefore, changes in telescope guiding or image seeing will cause slightly more or less source light to enter through the slit. These effects can cause large noticeable effects as well as more subtle possible unknown effects to be part of your final image. These issues, as well as items such as the use of a red sensitive slit viewing/guiding camera when observing very blue sources, will not be discussed further here (Wagner, 1992).

The angle on the sky of the spectrograph slit is very relevant to the use of CCDs as detectors. Observations of an astronomical source at some angle away from the zenith (an airmass of 1.0) can cause differential refraction to become an issue. Differential refraction is the variation of the angle of refraction of a light ray with zenith distance. All objects in the sky become slightly prismatic due to differential refraction by the Earth's atmosphere. If the spectrograph slit is not placed parallel to the direction of atmospheric dispersion, nonchromatically uniform light loss may occur at the slit since the image will have an extended, color-dependent PSF. Atmospheric dispersion is caused by the variation of the angle of refraction of a light ray as a function of its wavelength, and the direction parallel to this dispersion is called the parallactic angle. Using spherical trigonometry, the parallactic angle can be determined from the following:

$$\cos(\text{object declination}) \times \sin(\text{parallactic angle})$$
$$= \text{sign}(\text{hour angle}) \times \cos(\text{observer's latitude}) \times \sin(\text{object azimuth}),$$

where sign $= +1$ if the hour angle is positive and -1 if it is negative.

Table 6.2 presents an example of the amount of differential refraction that a point source will experience, as a function of airmass, for an observatory at an elevation of 2200 m. Note that all values in the table are relative to 5500 Å and are of significant magnitude at essentially all wavelengths and airmasses. CCD observations are particularly sensitive to an effect such as differential refraction as their very good red or blue QE could produce spectral data with highly inaccurate flux values. Even more troublesome is the case of using, for example, a blue flux standard star when observing red sources. Differential refraction can cause the light entering the slit, and therefore imaged by the CCD itself, to record incorrect intensities as a function of wavelength. Aligning the slit at the parallactic angle solves this problem.

A fun example to consider concerning differential refraction is the possible use of a direct CCD imager at very high airmass. The imager will act like a spectrograph and record spectra of sources using the Earth's atmosphere as the dispersive element. Some simple calculations will reveal that while possibly a good idea in principle, at the very high airmasses needed, the dispersion is extremely nonlinear and changes very rapidly with time. Extremely short exposures would be required, thereby allowing data to be gathered for only the brightest of stars. In addition, most astronomical telescopes can not point to the needed positions to obtain these sorts of very high airmass observations.

Some additional considerations related to the use of CCDs in astronomical spectrographs are those of order sorting filters, stray light, fringing, and economics. The first three of these cause undesired light to be imaged onto

Table 6.2. Example Differential Refraction Values (in arcsec)

Altitude	Airmass	3000 Å	3500 Å	4000 Å	4500 Å	5000 Å	5500 Å	6000 Å	6500 Å	7000 Å	8000 Å	9000 Å	10 000 Å
90	1.00	0.0	0.0	0.0	0.0	0.0	0.0	0.0	0.0	0.0	0.0	0.0	0.0
75	1.04	0.6	0.35	0.2	0.1	0.1	0.0	0.0	−0.1	−0.1	−0.1	−0.1	−0.2
60	1.15	1.3	0.8	0.5	0.2	0.1	0.0	−0.1	−0.1	−0.2	−0.3	−0.3	−0.4
45	1.41	2.3	1.45	0.8	0.4	0.2	0.0	−0.1	−0.2	−0.3	−0.5	−0.6	−0.6
30	1.99	4.0	2.45	1.4	0.7	0.3	0.0	−0.2	−0.4	−0.6	−0.8	−1.0	−1.1
20	2.90	6.3	3.9	2.3	1.2	0.5	0.0	−0.4	−0.6	−1.0	−1.3	−1.5	−1.7
15	3.82	8.5	5.3	3.1	1.6	0.7	0.0	−0.5	−0.9	−1.3	−1.7	−2.1	−2.3
10	5.60	12.9	8.0	4.7	2.4	1.1	0.0	−0.8	−1.3	−2.0	−2.6	−3.1	−3.5
5	10.21	26.1	16.2	9.5	4.9	2.2	0.0	−1.6	−2.7	−4.0	−5.2	−6.3	−7.0

the CCD, some of which falls directly on top of the object spectrum of interest. This extra light increases the apparent background, decreases the resulting S/N, and is often difficult or impossible to remove or even measure. As an example, using a diffraction grating to observe blue light in second order necessitates removal of the first-order red spectral light from the incident beam before detection by the CCD. Since CCDs are generally very red sensitive, this is a critical step to perform. A $CuSO_4$ filter is probably the best order sorting filter to use in this circumstance but is far from ideal. $CuSO_4$ filters have poor UV and blue transmission, are inconvenient to use (they are either a crystal or a liquid filter), and have a long wavelength red leak. Further discussion of such intricacies can be found in books concerning spectrographic observations of astronomical sources; also see Robinson (1988b), Pogge (1992), and Wagner (1992).

Finally, we conclude this section with our wish list for the ideal CCD to use within an astronomical spectrograph. The CCD should have low read and dark noise, high QE over a large wavelength range, very good CTE, small pixel size, and large dynamic range, and it should be "tuned" to the application desired. Tuning a CCD simply means that one should use the device that is best suited for the job at hand. Properties such as high red sensitivity, very deep pixel wells, small pixel size, back- side illumination and coating, etc. are items worthy of consideration.

6.4 Signal-to-noise calculations for spectroscopy

Calculation of the signal-to-noise ratio for spectroscopic observations is performed in a manner similar to that described earlier in this book (Chapter 4). As in photometric measures, we find that for bright sources $S/N \propto \sqrt{N}$ while for faint sources we must use the entire S/N expression (see Section 4.4). For spectroscopic observations obtained with a good CCD system, the largest noise contributors that will degrade the resulting S/N are the background sky contamination and how well the data can be flat fielded. The value of the S/N of a spectroscopic observation can have a few different meanings.[1] For CCD spectroscopy, one can calculate the S/N for the continuum or the S/N for a given spectral line.

---

[1] Note here that this is also true for CCD imaging or photometric applications. Depending on the user's choice of parameters such as n_{pix}, the final S/N value will change. Likewise, reported S/N values without comment on the exact choice of specific parameters are, at times, difficult to interpret.

In the continuum case, the number of pixels, n_{pix}, used in the S/N calculation will be determined by the continuum band-pass range over which the S/N is desired times the finite width of the spectrum on the CCD. For example, a typical CCD spectrograph might have an image scale of 0.85 Å/pixel and the imaged spectrum may have a width of 3 pixels on the array in the direction perpendicular to the dispersion. To calculate the S/N for the spectral continuum over a 100 Å band-pass in this example, one would use the value of 353 for n_{pix}. In contrast, a narrow emission line with a full-width at zero intensity (FWZI) of 40 Å would use an n_{pix} of 141. The S/N of the emission line will therefore be higher in value due to the smaller overall error contribution (approximately 3 to 1), not to mention the higher flux values per pixel within the line itself. As an exercise, the reader might consider how one would calculate the S/N in the case of an absorption line.

Signal-to-noise calculations are also useful in predicting observational values such as the integration time needed to obtain a desired scientific result from your spectroscopic observations. This type of calculation can be performed using the formulae presented in Section 4.4. Spectroscopic S/N values in the continuum of near 10 are often sufficient for gross spectral classification, whereas values in excess of 500 are needed for detailed abundance analysis or weak absorption line measurements. When making predictions of the S/N to expect in a spectroscopic observation, keep in mind that spectrographs are much less efficient overall than direct imaging cameras (2–4% vs. 30–40%) and that seeing effects and slit losses can be considerable in terms of the actual flux collected by the spectrograph.

6.5 Data reduction for CCD spectroscopy

This section discusses the basics of astronomical CCD spectroscopic data reduction. We start with Figure 6.6, which presents a raw, unprocessed 2-D CCD spectroscopic image of a point source. This figure illustrates a number of the observational and reduction items we discuss below. The initial reduction steps for CCD spectroscopy are exactly the same as previously discussed for imaging applications. Bias (or dark) frames and flat field calibration images are needed and used in the same way. After performing these basic reduction procedures for the CCD images, there are additional steps one must take that are specifically related to spectroscopy. These extra steps involve the use of spectra of spectrophotometric flux standards and wavelength calibration (arc) lamps. We will not discuss some minor, yet important, processing steps such as cosmic ray removal, bad pixel replacement, and night sky emission line

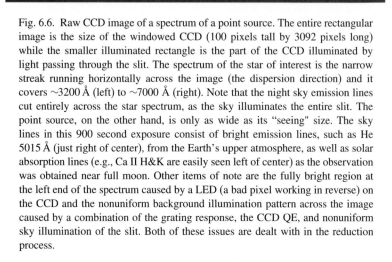

Fig. 6.6. Raw CCD image of a spectrum of a point source. The entire rectangular image is the size of the windowed CCD (100 pixels tall by 3092 pixels long) while the smaller illuminated rectangle is the part of the CCD illuminated by light passing through the slit. The spectrum of the star of interest is the narrow streak running horizontally across the image (the dispersion direction) and it covers ~3200 Å (left) to ~7000 Å (right). Note that the night sky emission lines cut entirely across the star spectrum, as the sky illuminates the entire slit. The point source, on the other hand, is only as wide as its "seeing" size. The sky lines in this 900 second exposure consist of bright emission lines, such as He 5015 Å (just right of center), from the Earth's upper atmosphere, as well as solar absorption lines (e.g., Ca II H&K are easily seen left of center) as the observation was obtained near full moon. Other items of note are the fully bright region at the left end of the spectrum caused by a LED (a bad pixel working in reverse) on the CCD and the nonuniform background illumination pattern across the image caused by a combination of the grating response, the CCD QE, and nonuniform sky illumination of the slit. Both of these issues are dealt with in the reduction process.

complexities. Detailed instructions for CCD spectroscopic data reduction can be found in Massey, Valdes, & Barnes (1992) and Pogge (1992). The article by Wagner (1992) is particularly useful, being the best review of the subject to date.

Figure 6.7 illustrates the five types of CCD spectral image needed for complete reduction and calibration of spectroscopic observations. The images shown in Figure 6.7 were all obtained with the same CCD, spectrograph grating, and telescope on two nights in June 2004. The spectral dispersion is 1 Å per pixel in the blue (1st order) and 2 Å per pixel in the red (2nd order) and the dispersion runs from red to blue (left to right).[1] Some items of note are 1) the bias level underlying the FeAr arc exposure, 2) the width of the flat field exposures compared with the point source stars (note spectrum d has been rescaled here to show the fringing in the red (left) part of the spectrum and thus looks thinner (see Appendix C)), 3) the absorption and emission lines in the spectra, and 4) cosmic rays in image g. These images do not show the full range of wavelength coverage shown in Figure 6.6 as they were expanded for clarity.

---

[1] Raw spectra are often displayed at the telescope in the manner they are read out from the CCD. While we all might like spectra displayed as blue to red or increasing wavelength from left to right, at times, raw spectra are displayed opposite to that as shown here. One quickly develops a mental flip ability while observing or learns the needed plot command to display the spectrum "correctly."

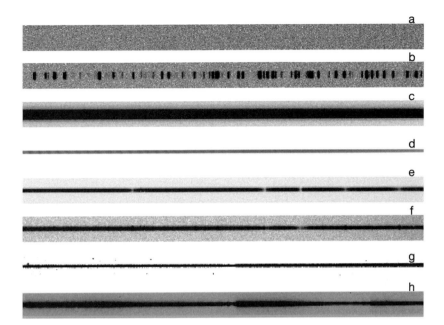

Fig. 6.7. The five necessary types of CCD image needed for spectral reduction. Image a is a bias frame, b is an FeAr calibration lamp (arc) exposure, c is a quartz lamp projector flat field in the blue spectral region, d is the same in the red region, e is a blue observation of the spectrophotometric flux standard star BD +26 2606, f is the same in the red region, and g and h are the stars of interest, SX Her and Z UMa respectively. See Figure 6.8.

Figure 6.8 shows line plots of each of the images in Figure 6.7. The bias frame shows that this CCD has an increase in noise by 10–15 ADU at the very end (left) of the readout near the output amplifier. This is fairly typical and of no concern as it covers only a few columns and is accounted for in the reduction process. The FeAr arc lamp shows nothing but emission lines whose known wavelengths are used to determine the CCD pixel number to reduced wavelength solution. We see that the blue quartz flat field, c, rises to the red (as quartz lamps are red, peaking redward of your eye's sensitivity) while the red flat field shows the characteristic fringing caused by the long wavelength red photons passing through the CCD, reflecting back, and interfering with themselves. This horrid looking fringe pattern is also accounted for and removed during reduction. The high count rate flat fields also show the typical spike at the ends of the readout (due to amplifier turn on/off) and the small overscan region (little flat portion at the bottom right of the plots) of the CCD. We also show the same standard star (BD +26 2606)

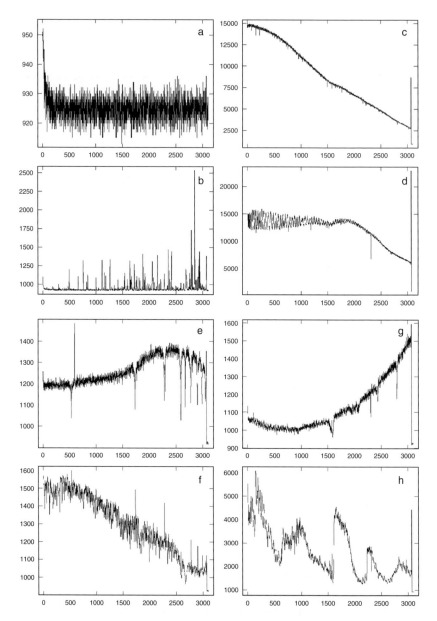

Fig. 6.8. Plots of the raw spectra shown in Figure 6.7. These plots show the full spectral range of the instrument (3096 pixels = ~1200 Å in the blue range and ~2400 Å in the red range – see Figure 6.9) and are plots of a single CCD row taken through the middle of the spectra in Figure 6.7. These raw data are plotted as pixel number (*x*-axis) vs. ADU (*y*-axis) as is typical, with red being to the left. Note the spectral shapes, which are dominated here by the source convolved with the grating response and the CCD QE. The conversions from pixel number and ADU to wavelength and flux (as well as a proper spectral shape) are done in the reduction process.

in the red and blue regions to illustrate its differences. The blue region (e) shows the Balmer absorption series in this sdF star and a cosmic ray (narrow emission spike near Hβ) while the red region shows the fairly large atmospheric absorption bands due to molecular oxygen (the A and B bands) and fringing. The two objects of interest are SX Her, a long period variable star with emission lines due to a recent shell ejection, and Z UMa, a cool supergiant with strong TiO absorption bands.

For a point source, the final product desired is a 1-D spectrum consisting of wavelength versus relative or absolute flux. It is assumed that you have in hand CCD spectra of your object(s) of interest, at least one observation of a flux standard, wavelength calibration spectra of arc lamps, and bias and flat frames (see Figure 6.7). The first step with the reduced two-dimensional CCD image, after the standard processing with bias and flats is performed, is to extract the spectrum itself and collapse it into a 1-D image of pixel number versus ADU/pixel. In the simplest (and unrealistic) case, the imaged spectrum lies exactly along one row (or column) of the CCD and one can extract it simply by extracting the single row from the final image. Generally, the imaged spectrum covers two or three or more rows (or columns) and the extraction process involves some manner of summing a few adjacent rows perpendicular to the dispersion (an "after-the-fact" pixel binning) at each point along the dispersion. Furthermore, in practice it is often found that CCD spectra are not precisely aligned with the CCD pixels and are curved on the detector as the result of the camera optics, instrumental distortions, or CCD flatness issues. Fainter sources present additional complexities, as centroiding the spectrum (by, for example, using cuts across the dispersion direction) in order to extract it is often difficult or impossible. A typical example might be a spectral observation of a faint continuum source that contains bright emission lines. Details of the spectral extraction process, sky subtraction, and optimal extraction techniques are discussed in Schectman & Hiltner (1976), Horne (1986), Robinson (1988b), Robinson (1988c), Pogge (1992), and Wagner (1992).

At this point, your extracted spectrum will have an x axis that is in pixels and we wish to convert this to wavelength. The procedure to perform such a task involves observations of calibration arc spectra obtained often during your observing run. The idea is to match the x-axis pixel scale of the calibration arc lamps with their known wavelengths and then apply this scaling procedure to your object data. Calibration arc spectra of sources such as hollow cathode Fe lamps or He-Ne-Ar lamps contain numerous narrow emission lines of known wavelength. Use of these emission lines during the reduction procedures allows a conversion from a pixel scale to a (possibly rebinned linear) wavelength scale. Correction for atmospheric extinction (Hendon & Kaitchuck, 1982), similar in manner to photometric corrections already discussed but generally a bit more complex due to the larger wavelength coverage, can now be applied to all obtained spectra. Since you are relying in this step on the hope that the arc emission lines fall onto specific CCD pixels and define a wavelength scale that is identical to that of your object, you want to obtain arc data in as similar a manner as possible to that used to gather your object spectra. Instrument flexure caused by telescope motion throughout the night or movement of the CCD within the dewar are two of many possible effects that will invalidate the wavelength to pixel scaling procedure.

We now have to deal with our collected spectra of flux standards. These stars are observed solely for the purpose of converting the collected pixel count values into absolute or relative flux values. Application of all of the above steps to your spectra of spectrophotometric flux standards will produce 1-D data with a wavelength scale (x axis) versus counts on the y axis. We now wish to have the y axis of ADUs or counts converted into flux units such as ergs s^{-1} cm^{-2}Å^{-1}. Most observatories and data reduction packages (such as IRAF and MIDAS) contain lists of spectrophotometric flux standards appropriate to observe and use for fluxing of your spectroscopic data. Within the reduction software, tables of wavelength versus flux are kept for a large number of spectrophotometric flux standards. To understand and appreciate the details involved in setting up even a single spectrophotometric standard star, see Tüg, White, & Lockwood (1977), Jacoby, Hunter, & Christian (1984), and Massey *et al.* (1988). In a similar manner to the method by which we took the known arc wavelengths and converted their pixel scale into a wavelength scale, we can now take the known fluxes of the standard stars and convert pixel counts into relative or absolute fluxes. The difference between relative and absolute is essentially the difference between a narrow or large slit width as mentioned above. The conversion of counts to flux is performed under the assumption that slit losses, color terms, transparency, and seeing were similar between the standard star observations and the object(s) of interest.

One can never have too many calibration data and must always trade off time spent collecting CCD frames of standard stars and arcs with collection of data for the objects of interest. Instrument flexure, nonphotometric conditions, color terms, and accurate wavelength calibration are crucial to the production of accurate final spectroscopic results such as that shown in Figure 6.9.

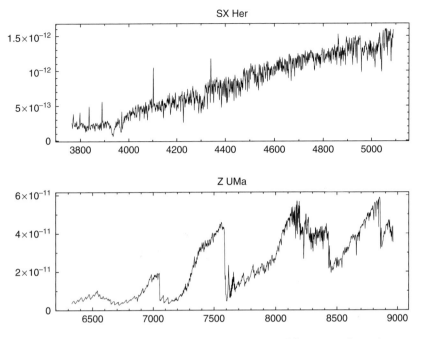

Fig. 6.9. Final reduced CCD spectra of the two stars of interest as shown in Figure 6.8. We see that the *y*-axis values of counts in ADU have been transformed into flux in units of ergs s^{-1} cm^{-2}Å^{-1}, using the standard star observations. The *x*-axis units are now wavelength in Å increasing to the right. Narrow emission lines of the Balmer series as well as Ca II H&K absorption are seen in SX Her and Z UMa's spectrum is dominated by TiO absorption bands.

6.6 Extended object spectroscopy

The definition of extended object spectroscopy follows from the fact that we wish to obtain spectra, not from an unresolved point source, but from a region of the sky for which we desire simultaneous wavelength and spatial information. Examples might include galaxies, nebulae, and planets within our solar system. While there is no fundamental difference between this type of spectroscopy and point source observations such as those described above, there are differences in the instruments used and the reduction techniques involved. We will present here a very basic introduction to the subject and refer the reader to the more detailed review given by Pogge (1992).

One method of obtaining spectra of an extended object is by using long-slit spectroscopy. While sounding like an entirely new method of observing, long-slit spectroscopy is very basic. When we discussed point source observations,

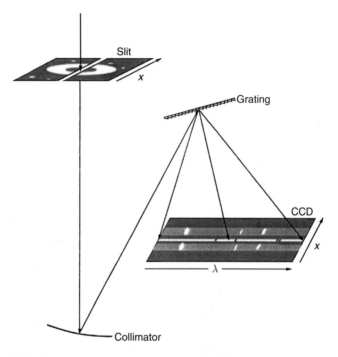

Fig. 6.10. Schematic view of a typical long-slit CCD spectrograph. Positions along the slit are mapped in a one-to-one manner onto the CCD detector. A number of optical elements in the camera, used to re-image and focus the spectrum, have been omitted from this drawing. From Pogge (1992).

we concerned ourselves with the details of the spectrograph and final spectrum as they related to the incident light from a point source focused on the spectrograph slit and imaged on the CCD array. We mentioned in that discussion that we desired to use the slit to keep out as much background light as possible. However, all objects that lie along the slit, say in the x direction as presented in Figure 6.10, will produce a spectrum and be imaged on the 2-D CCD array. This last statement makes some assumptions about the ability of the spectrograph optics and optical path to support such an endeavor.

One can imagine a case in which a number of point sources are lined up in an exact east–west manner and a spectroscopic observation is performed of these stars using an east–west slit alignment. The output image will contain a number of parallel spectra, one for each point source, lined up on the CCD. Obvious extensions of this simple example to real cases include placing the long slit of a CCD spectrograph along the axis of a spiral galaxy or alternately across and alongside of bright knots of ejecta within a recently erupted classical nova. The uses of such spectrographic techniques are many and yield information on both the spectral properties of an object (radial velocity, line widths, etc.) and the spatial properties (spiral arms versus the bulge, etc.). This type of observational program is indeed another that benefits from large-format CCDs.

A more complex and upcoming type of two-dimensional CCD spectrograph is the imaging Fabry–Perot interferometer. Within this type of a spectrograph, 2-D spectral imaging over a narrow band-pass can be collected on the CCD, and with a change in the Fabry–Perot etalon spacing, a new narrow band-pass can be isolated. These types of instruments are often quite complex to master and setup, produce complex data cubes of resultant imagery, and tend to be less efficient than conventional spectrographs due to their many optical surfaces. However, to its benefit, Fabry–Perot interferometry has a unique capability to provide tremendous amounts of scientific data per observational data set and to obtain fantastic results for the objects observed. The details of this type of spectrograph and its use can be found in Roesler *et al.* (1982), Bland & Tully (1989), and Pogge (1992).

A final method of obtaining spatial as well as spectral information is that of narrow-band imaging. The spatial aspect comes about through the use of a CCD as a wide-field imager and the spectroscopic information is provided at selected wavelengths by the use of narrow-band filters. This is no different from observing a given area of the sky in B, V, and R filters, except that here one usually is imaging one or more extended objects and selects carefully two or more restrictive filters in order to obtain specific results. Simultaneous imaging in all colors is lost as one has only limited spectral coverage with this technique, but one obtains full spatial sampling over the entire field of view and the instrumental setup and data reduction are quite simple. In addition, this type of 2-D spectrophotometry can be performed with any CCD imager with the addition of the appropriate filters.

There are numerous other complexities associated with extended object spectroscopy, some in the observational setup procedures needed and some in the data handling and reduction procedures. However, the wealth of information available from such observations far outweighs the difficulties of these more complex types of CCD spectroscopy.

6.7 Slitless spectroscopy

The above discussion of astronomical spectroscopy was based on the use of a spectrograph that allowed an image of a slit to be dispersed and then re-imaged onto the CCD detector. Another method, slitless spectroscopy, has also been a workhorse for astrophysical science (Bowen, 1960a) and continues to be useful today (MacConnell, 1995). The idea of slitless spectroscopy is to take advantage of a simple direct imaging system to provide a mechanism for producing and recording astronomical spectra. Some form of dispersive element is placed either before the telescope entrance aperture (e.g., an objective prism) or just before the detector (e.g., a grism), allowing a spectrum to be formed and imaged on the detector.

Work with objective prisms placed in the optical path of a Schmidt telescope and the resulting slitless spectra being imaged on large-format photographic plates has been used in astronomy for over seventy years. This type of an arrangement allows one to obtain spectra of all imaged objects in the field of view with the integration time setting the limit for the faintest usable spectra (Walker, 1987). The resulting spectra are generally of low dispersion and used for gross classification purposes (MacConnell, 1995), but even radial velocities can be measured (Fehrenbach, 1967). The dispersions available with an objective prism arrangement are near 200–300 Å/mm and the spectra obtained generally cover a total band-pass of 400–1000 Å. The limiting magnitude for a usable spectrum with an objective prism setup (for a 1-m, f/2 telescope and a 1 hour integration) is approximately given by

$$m_{\mathrm{lim}} = 18.5 + 5\log_{10} T,$$

where T is the focal length of the telescope in meters. Generally, this limit (here equal to about 20th magnitude) is 5 to 6 magnitudes brighter than that available at the same integration time with the same optical/detector system when used as a direct CCD imager.

Objective prisms have the disadvantage of being massive and producing nonlinear dispersions. Transmission gratings have replaced objective prisms in some applications although they can introduce coma into the final image. To counter the effect of coma, grating-prism combinations or grisms were developed (Bowen, 1960b). A variant of this idea is the grens or a grating-lens combination. Gratings can be blazed to produce high throughputs in a single order, but the zeroth-order image is always present and leads to field crowding, background light issues, and possible confusion as an emission line in another spectrum (Schmidt, Schneider, & Gunn, 1986).

Current work with slitless spectroscopy of interest here are those applications that use a CCD as a detector. Such examples are the PFUEI (Schmidt, Schneider, & Gunn, 1986; Gunn & Westphal, 1981) and COSMIC (Kells *et al.*, 1998) instruments used at the 200″ Hale telescope and the QUEST instrument (Sabby, Coppi, & Oemler, 1998) used on a 1-m Schmidt telescope.

These instruments use a transmission grating, a grism (preceded by a multi-slit aperture mask), and an objective prism respectively to form their spectra, which are then imaged onto a CCD. The above formula for the limiting magnitude of a useful spectrum should be revised fainter by 1–2 magnitudes for a high throughput, high QE, low noise CCD slitless spectroscopy system.

All types of slitless spectroscopy must cope with three similar issues in terms of their image acquisition and data reduction. First, it is often useful to obtain a direct image of the field without the dispersing element in place. This image allows accurate source location independent of intervening spectral overlap or zero-order confusion (in the case of using a grating). Slitless spectra can cover many hundreds of pixels on the CCD with the zeroth-order image being separated by 100–200 pixels. Second, data extraction from the final image can often be complex as field crowding or seeing changes during an exposure will cause broadened, ill-defined spectra. Procedures here generally use a rectangular extraction box containing the spectrum, yet being small enough to avoid large sky contributions. Finally, there is the issue of calibration of the spectrum in flux and wavelength. Wavelength calibration is usually accomplished by one of two methods: observation of an astronomical source with well-defined, known spectral (usually emission) lines or the use of a calibration lamp shone through pinholes placed at the focal plane. Other methods include use of known spectral lines in one or more of the imaged spectra or the centroid of the zeroth-order image[1] and a knowledge of the plate scale to calculate the dispersion per CCD pixel (Schmidt, Schneider, & Gunn, 1986).

Of the above three concerns, the calibration of the obtained spectra is the most important and the hardest to perform. It is often the case that slitless spectra are presented as unfluxed, having a spectral shape dominated by the instrument and detector response, and often with a wavelength scale that is only approximate. The latter is usually sufficient, as dispersions of 10–50 Å/pixel or more provide wavelength resolutions of only 10 to 200 Å or so. Thus, precise wavelengths are often unimportant. Providing fluxes for the spectra is a difficult and often ignored issue (see below), although not always (Schmidt, Schneider, & Gunn, 1986; Schmidt & Gunn, 1986). Spectral classification, identification of (redshifted) emission lines, or separation into blue and red objects are typical goals, none of which require more than relative flux values.

---

[1] The zeroth-order image, while possibly a nuisance, can be used to obtain photometric information such as an instrumental magnitude or, for time-series slitless spectroscopy, a light curve.

Two considerations in the determination of the flux from an object imaged via slitless spectroscopy are background subtraction and image flat fielding. The determination of the sky background in a CCD image of slitless spectra can be confusing. While all the light from the astronomical objects is passed through some sort of dispersing element, so is the light from the background sky. Thus, at any point on the CCD detector, the background is a combination of spectrally dispersed but unresolved sky light. Thus, a "sky" section of the CCD lying adjacent to an object spectrum is not a true representation of the (dispersed) background level imaged with the spectrum. Sky subtraction is not a simple procedure and is often not performed.

Flat fielding a slitless image also presents challenges. We discussed above how the color terms in a flat field image can have large effects on the outcome of the calibration procedure. You can imagine that color terms are even more important in this circumstance, as the flat field light is imaged and dispersed across the CCD in a hopefully similar manner to that of your data. As expected, sky flats provide a far better flat field than dome flats (Schmidt, Schneider, & Gunn, 1986).

Let us examine a few slitless spectroscopy systems in detail. The prime focus universal extragalactic instrument (PFUEI) (Gunn & Westphal, 1981; Schmidt & Gunn, 1986; Schmidt, Schneider, & Gunn, 1986) used a TI CCD with 15-micron pixels, an image or plate scale of 0.4 arcsec/pixel, and a field of view of 30 arcmin. A transmission grating with 75 lines per mm, blazed at 6500 Å, is used to provide a spectral dispersion of 35 Å/pixel. The transmission grating can be easily removed from the optical path allowing a direct CCD image to be obtained. When placed in the beam, spectra of each object within the field of view are recorded on the CCD. Figure 6.11 shows a portion of a typical slitless spectroscopic CCD image obtained with the PFUEI.

An innovative setup can combine CCD drift scanning techniques with slitless spectroscopy. The QUEST instrument (Sabby, Coppi, & Oemler (1998) and Section 4.6.4) is such a camera operating on a 1-m Schmidt telescope and using a CCD array of sixteen 2048×2048 Loral CCDs. A 3.2° objective prism is employed to provide 401 Å/mm dispersion, yielding a spectral resolution of $\sim$10 Å. A cutoff filter is used to stop light longward of 7000 Å. This cutoff filter reduces the increased redward sky background as well as shortening the length of the imaged spectra, thus decreasing the chance of spectral overlap. Since a prism is used as the dispersing element, no zero-order image concerns exist; however, the spectra do have a nonlinear dispersion.

Figure 6.12 shows a portion of a QUEST drift scan with the prism in place. The spectra are about 400 pixels long and have the dispersion direction

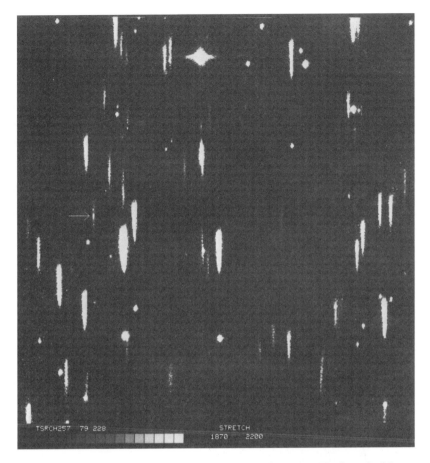

Fig. 6.11. A portion of a PFUEI spectroscopic CCD image. The figure is 5.5 arcmin on a side with east at the top and north to the left. Note the zeroth-order (point-like) images for each spectrum that appear about one spectral length to their west. The large star-like object near the top middle of the frame is a saturated zeroth-order image from an object whose spectrum is off the top of the image. The zeroth-order images are at the positions of the sources as seen in a direct image and are used in the PFUEI data reduction to set the wavelength scale. The arrow points to the C IV 1550 Å line in a redshifted 19.5 magnitude quasar. From Schmidt & Gunn (1986).

aligned with the scan direction to avoid spectral clipping at the CCD edges. No attempt is made to flux calibrate the data but initial wavelength calibration is accomplished by identification of known spectral lines detectable in stars.

Slitless spectroscopy is a wonderful tool for obtaining spectra of many objects simultaneously and to very faint limits. The simplistic nature of the

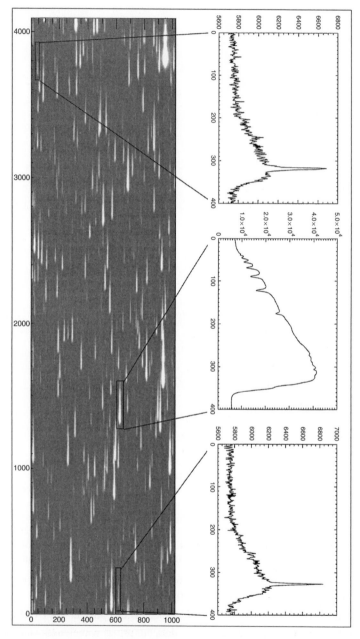

Fig. 6.12. A 1000 × 4000 pixel section of a QUEST CCD drift scan obtained with the use of an objective prism. The dispersion is parallel to the scan direction and each spectrum is 200–400 pixels in extent. Examples of some extracted spectra are shown on the right. From Sabby, Coppi, & Oemler (1998).

technique makes for an ease of use given that a direct imaging CCD camera is available. The addition of slitless spectroscopy with large-format CCDs and CCD drift scanning go even further, allowing spectroscopic identification of many thousands of objects with little additional effort beyond the direct imaging already in place. Slitless spectroscopy promises to be an evolving CCD application, yet to have reached its peak potential.

6.8 Exercises

1. Compare spectroscopic results obtained under similar conditions with the same telescope if using a SITe CCD or a LBL CCD as described in Table 3.2. What are the most important differences?
2. What is the difference between, and the reasons for, obtaining a quartz lamp exposure and an arc lamp exposure when performing spectroscopy with a CCD?
3. Derive the equation given for the projected slit width in Section 6.1.
4. What value for R, the spectral resolution, is needed to allow separation of the two forbidden oxygen III lines at 4959 Å and 5007 Å?
5. Describe briefly, the differences and similarities of point source spectroscopy, multi-fiber spectroscopy, echelle spectroscopy, fiber spectroscopy, and slitless spectroscopy. Name one or two observational programs that can be best performed using each method.
6. Discuss the advantages of a double spectrograph. Can you design a single spectrograph that uses a reflection grating as the dispersive element, which would allow the entire optical spectral range (3000–10 000 Å) to be imaged on to a CCD at once? What if a prism were used instead of a grating?
7. How can poor charge transfer efficiency effect spectroscopic observations with a CCD? Discuss this answer for the cases of sources with strong emission lines, strong absorption lines, and no spectral lines at all.
8. Produce a flow chart of a typical reduction procedure for spectroscopic CCD observations. Clearly show which types of calibration images are needed and when they enter in to the reduction process.
9. What type of star is best used for a spectrophotometric standard star? Discuss the specific properties of stellar brightness and spectral features in the star's spectrum.
10. Provide a quantitative discussion of why a narrow slit is used for radial velocity work but a wide slit is used for spectrophotometry.

11. Derive the formula given (in Section 6.3) for the relation between the parallactic angle and the position of an astronomical object.

12. Design an observing plan that would make use of the Earth's atmosphere as the dispersive medium to allow spectrographic observations to be made with a CCD imager. (Note: the data given in Table 6.2 may be of use here.) Give specific details of integration times needed and the pixel extent of the recorded spectra. What are the most serious flaws in such a plan?

13. Repeat the sample S/N calculation given in Section 6.4 for a spectroscopic observation made with a CCD. What happens if the sky brightness is doubled or tripled? How does the S/N depend on wavelength?

14. Why is sky subtraction so important for CCD spectroscopy? Was it important for spectroscopic data obtained before CCDs were used?

15. Design an observational project that would greatly benefit from the use of a two-dimensional imaging Fabry–Perot interferometer.

16. What is the difference between a prism, a grism, a transmission grating, and a grens? Find a working instrument that uses one of these devices and discuss its performance.

17. Describe in detail, the differences and similarities of the PFUEI and the QUEST instruments. Which is a better choice to use to obtain spectra for a sample of very faint sources? Which is best for a large area spectroscopic survey?

7

CCDs used in space and at short wavelengths

The current high level of understanding of CCDs in terms of their manufacture, inherent characteristics, instrumental capabilities, and data analysis
techniques make these devices desirable for use in spacecraft and satellite
observatories and at wavelengths other than the optical. Silicon provides at
least some response to photons over the large wavelength range from about
1 to 10 000 Å. Figure 7.1 shows this response by presenting the absorption
depth of silicon over an expanded wavelength range. Unless aided in some
manner, the intrinsic properties of silicon over the UV and EUV spectral
range (1000–3000 Å) are such that the QE of the device at these wavelengths
is typically only a few percent or less. This low QE value is due to the
fact that for these very short wavelengths, the absorption depth of silicon is
near 30–50 Å, far less than the wavelength of the incident light itself. Thus,
the majority of the light ($\sim 70\%$) is reflected with the remaining percentage
passing directly through the CCD unhindered.

Observations at wavelengths shorter than about 3000 Å involve additional
complexities not encountered with ground-based optical observations. Access
to these short wavelengths can only be obtained via space-based telescopes or
high altitude rocket and balloon flights. The latter are of short duration from
only a few hours up to possibly hundreds of days and use newly developing high-altitude ultra-long duration balloon flight technologies. Space-based
observations in the high energy regime from UV to shorter wavelengths usually require detectors to be "solar blind." The term solar blind means that the
detector must be completely insensitive to visible light photons. This is generally accomplished by using a non-optically active type of detector or through
the use of various types of filters. The majority of astronomical objects emit
10^4–10^6 visible light photons for every UV or shorter wavelength photon;
thus even a visible light blocking filter with a 1% visible transmission is not
nearly sufficient to remove optical contamination. In addition, most common

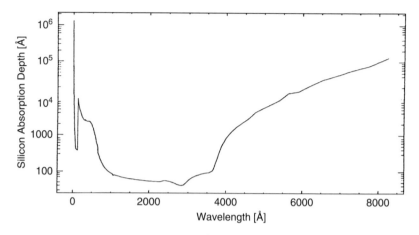

Fig. 7.1. Silicon absorption depth (in Å) from 1.2 to 8500 Å. The vertical axis is a log scale with major tick marks starting at 100 and ending at 10^6. From Bonanno (1995).

filters used to block visible light also absorb some of the incident higher energy radiation as well. Use of such absorbing filters causes even a high QE CCD at UV wavelengths (say 20%) to be reduced to a low effective QE near 2%.

Long-term exposure to high vacuum can cause contamination of the dewar in which the CCD resides. This contamination can be through outgassing of various materials such as vacuum grease or AR coatings (Plucinsky *et al.*, 2004), normally not a problem in well-produced ground-based systems. Exposure to high energy radiation can cause changes in the QE of the CCD or cause permanent damage to the pixels and electronic structures within the array. Studies of the effects of high energy radiation and space environment observations on CCDs are ongoing at a number of laboratories such as the Jet Propulsion Laboratory (for NASA space-based satellites and missions), the Space Telescope Science Institute, and at the European Space Agency (ESA). Good discussions of space-based CCD usage are presented in Janesick, Hynecek, & Blouke, 1981; Janesick, Elliott, & Pool, 1988; Holtzman, 1990; Janesick & Elliott, 1992; Janesick, 2001; Strueder *et al.*, 2002; Meidinger *et al.*, 2004a; Meidinger *et al.*, 2004c, and a number of the websites listed in Appendix B.

Before we discuss the details of observations at wavelengths shorter than the optical, we need to make a brief detour to look into some special issues related to space-based observations with CCDs. The more notable of these issues are the calibration of the CCD throughout the instrument or mission lifetime, the fact that the point-spread function is much smaller than generally obtained with ground-based data, and continual degradation of the CCD with time as the result of radiation damage.

7.1 CCDs in space

Space-based CCDs have a number of special problems associated with them that are often not considered for ground-based systems. Once launched, human intervention is unlikely and the CCD and instrument package can never be retrieved for fault correction or calibration purposes. Even simple procedures, such as bias calibration, take on new meaning as CCD evolution or changes in the gain or other CCD electronics mean new calibration images are needed. Damage to the array (see Section 7.2), or the possibility that the primary circuits fail and the backup electronics or even a different clocking scheme must be used, means that new calibration images must be produced. Also, each observer does not have the ability to obtain all the needed calibration data and the project must provide the finest and most up-to-date calibration images for each CCD, instrument, and mode of operation. All issues have to be thought out completely prior to launch or dealt with through analysis of downloaded data during the mission.

One such example of a significant change in CCD operation is provided by the Hubble WFPC2 instrument (Holtzman *et al.*, 1995b). After operating in space for only about three months, it was noticed that the CCDs developed an odd sort of CTE effect. The effect caused stars to appear fainter if imaged in higher numbered rows. The apparent cause was the development of a large number of traps within the CCDs not seen during preflight tests. Photometric gradients of 10–15% were present along CCD columns and, even worse, the effect was highly dependent on the brightness of the imaged star, being only about 5% for bright stars.

Using ground-based laboratory tests with similar electronics and CCDs, it was determined that changing the operating temperature from $-76°$ C to $-88°$ C would cause a sharp decrease in the CTE effect. Such a change caused the CTE variations to almost disappear, leaving only a 3–4% gradient. A further temperature decrease would probably improve the situation but in-flight hardware did not allow the CCDs to be operated at colder levels. Thus considerable effort has been put into the development of a semi-empirical software model that can be applied to data obtained with the WFPC2 in order to correct for the remaining effect (Holtzman *et al.*, 1995a; Whitmore & Heyer, 1998). A number of the CCDs in HST instruments and those in the

Chandra X-ray observatory have shown long-term degradation in their CTE performance due to the radiation environment of the telescopes' orbit. For example, the STIS CCDs have changed from a CTE of 0.999 999 to 0.999 91 since launch (Kimble *et al.*, 2000).

One consequence of the CCD operating temperature being lowered in the WFPC2 was decreased dark current. However, on-orbit hot pixel development was greater than expected with many of these hot pixels "fixing" themselves after dewar warming (Section 7.2). Calibration dark frames are therefore required often to monitor the dark current and to provide the best dark frames to use given any set of observational circumstances. Hot pixels are especially important to understand in space-based CCD imagery as the very small PSF of imaged scenes and the appearance of numerous cosmic rays with a plethora of shapes, including single pixel events, must be distinguished from the collected flux of interest.

We alluded above to the importance of cosmic ray identification in order to avoid misinterpretation of imaged scenes. From a sample of 2000-second dark images taken with the WFPC2 it was found that 5–10% of the cosmic ray events were single pixel events of 5 sigma or greater above the bias level. Fully one half or more of these events showed consistent pixel positions from frame to frame and thus could not be identified with true cosmic rays or local radioactivity from the dewar and surroundings. Typical signal levels for true single pixel cosmic ray events were near 200 electrons while multiple events peaked near 700 electrons (Holtzman *et al.*, 1995b). Multiple pixel cosmic ray hits (averaging seven affected pixels per event) are much more common than single pixel events, and a rate of almost two events per CCD per second was observed.

CCD dewars, once sealed, evacuated, and chilled, are often seen to produce contaminants owing to outgassing of grease or other coatings used in their construction. When at operating temperatures of $-80°\,C$ or so, the dewar window is a good site for condensation of such contaminants. These small particles of material are very good absorbers of light, particularly UV and visible radiation, because of their characteristic sizes. A likely cause of the contamination is C, O, and F atoms that often form a thin layer on the dewar window or instrument filters quickly and then increase this layer slowly with time. Bake-out procedures have been modeled as a possible method to reduce the thickness of these layers (Plucinsky *et al.*, 2004) specific to the ACIS CCDs on the Chandra X-ray observatory.

One simple calibration test that allows monitoring of this effect is to obtain fairly regular observations of a bright UV star. If the dewar window does indeed get fogged with material, careful measurements of the UV throughput of the observed flux will show a slow degradation. Even in the best space-based instruments, small amounts of material outgas, and after several weeks UV performance can be noticeably lower. One solution that seems to work, at least for the Hubble Space Telescope WFPC2 CCDs and the Chandra X-ray observatory ACIS CCD imager (and for general observatory dewars), is to warm the dewar up to allow for thermal desorption. The WFPC2 CCDs were warmed to near 20° C for about 6 hours approximately every month. In a typical observatory dewar after warm up, one can attach a vacuum pump and pump out the now non-frozen water and other contaminants, then recool the device.

Flat fields, as we have discussed before, are very important to have in one's calibration toolkit. Once in orbit, either as a satellite or as a spacecraft destined for another world, the CCDs aboard generally have little ability to obtain flat field calibration images. High S/N flats made prior to launch in the laboratory are often the best available. These usually provide overall correction to 5% or a bit better, but small effects, such as illumination or instrument changes, limit the accuracy of the correction. Sometimes, the space-based CCD has significant changes, and large corrections are needed or new flats have to be generated in some manner.

The original WFPC camera aboard Hubble could obtain on-orbit flats through observation of the bright earth (Holtzman, 1990; Faber & Westphal, 1991). These were not elegant flats, having streaks and nonuniformities, but were all that was available. WFPC2 used Loral CCDs, which have an increased stability over the original TI CCDs, allowing preflight laboratory flats to work very well, even after the reduction in operating temperature as discussed above. Numerous other small effects, such as color dependence, radiation damage, hot pixels, CCD illumination, and optical distortions seen in the on-orbit WFPC2 flats are discussed in detail in Holtzman *et al.* (1995b). The effects of flat fielding, CTE, and the other issues discussed above on the photometric performance of the Hubble WFPC2 are described in Faber & Westphal (1991), Holtzman *et al.* (1995a), and Whitmore & Heyer (1998).

The Galileo spacecraft certainly provided impressive imagery of the planet Jupiter and its satellites and was one of the first public CCD cameras to be launched into space. Its CCD camera is described in detail in Belton *et al.* (1992) and can be used as an example of the details of space-based observations, their calibrations, properties, and difficulties. CCD and instrument stability and processes for their calibration after launch are major effects to consider as well as proper treatment of the photometric calibration images in lieu of the much reduced PSF.

The solid-state imager (SSI) aboard Galileo consisted of a single 800×800 TI CCD with a read noise of 40 electrons, gains of 38 to 380 electrons per DN, and a pixel size of 15 microns yielding 2.1 arcsec per pixel. The SSI, like the WFPC2, developed a CTE problem after about 8–12 months in space. Detailed study of SSI images taken during periods of cruise science (Howell & Merline, 1991) revealed that the CTE problem resulted in a readout tail containing 400 electrons, independent of the brightness of an imaged star or its location within the CCD. The cause was attributed to a trap, not in the active CCD array, but in the output register. Radiation damage (see next section) was the most likely cause. Due to the constant number of trapped electrons, photometric correction was possible to a high degree of accuracy.

Point sources imaged in space are free from the blurring effects of the Earth's atmosphere and have a very small PSF compared with those commonly obtained with ground-based telescopes. A theoretical diffraction-limited image formed through a circular open aperture will have a FWHM (of the Airy disk) in radians of

$$\text{FWHM} = \frac{1.03\lambda}{D},$$

where λ is the wavelength of observation and D is the diameter of the aperture (Born & Wolf, 1959). Note that if we were to use the radius of the first Airy disk dark ring as our definition of image size, we would have the traditional formula

$$r = \frac{1.22\lambda}{D}.$$

Figure 7.2 shows theoretical Airy disk PSFs expected to be imaged by the SSI at three representative wavelengths and five different possible slight de-focus values.

The FWHM of the SSI images (being obtained without any atmospheric or other seeing effects) were predicted to be about 0.55 arcsec at 4000 Å and 1.2 arcsec at 9000 Å. These PSF sizes correspond to 0.25 and 0.6 pixels respectively, making the SSI images severely undersampled ($r \sim 0.2$). This level of undersampling makes it impossible to directly determine the true FWHM or profile shape of a PSF. Using multiple images with slight offsets, images containing multiple stars with different pixel grid placements, and model CCD images, one can reconstruct the true PSF imaged by an under-sampled space-based CCD camera. In the SSI case, the PSF was found to be slightly larger than predicted and attributed to a slight camera focus problem.

As we have seen, undersampled images will lead to astrometric and pho-tometric error, as the lack of a well-sampled PSF makes it hard to determine the true image center or the actual flux contained in the image. For the SSI,

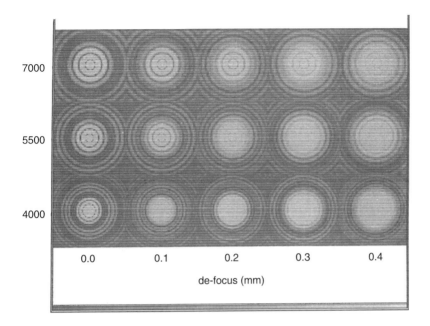

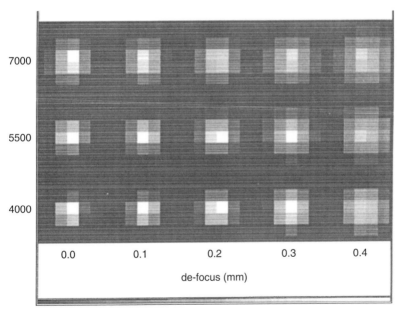

Fig. 7.2. Modeled Airy disk patterns imaged by the Galileo SSI. The top panel shows the calculated PSFs as would be seen under very well-sampled conditions while the bottom panel shows the same PSFs as they would appear when imaged by the SSI. The severe pixelization of the PSFs is apparent. The rows are for 7000, 5500, and 4000 Å (top to bottom) and the five columns are (left to right) de-focus values for the SSI camera in mm. From Howell & Merline (1991).

astrometric error amounted to about 0.8 arcsec even for bright stars, or about half a pixel. Observations of bright guide stars are a common occurrence for spacecraft and are used for navigation and course correction. Large astrometric uncertainties are hazardous and can lead to spacecraft orbital trajectories with inaccurate pointings, having the potential of producing spacecraft course corrections that could cause it to miss a target or, even worse, come too close. In the Galileo case, it was determined that a large number of guide star images was needed and careful analysis of these could be used to determine the path and navigation of the spacecraft within acceptable limits.

Photometrically, the nature of the undersampling manifests itself in two ways. First is the way in which one extracts the data and how a flux value is assigned to it; second is the effect of digitization noise, which is large for the SSI. Figure 7.3 illustrates the first of these issues by presenting SSI data for a bright star. Because of the nature of the PSFs imaged with the SSI,

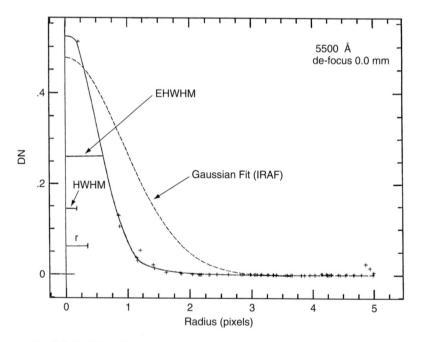

Fig. 7.3. Radial profile plot of a bright star imaged by the Galileo SSI. The plus signs are the actual CCD DN values (for $G = 380\,e^-/\mathrm{DN}$) and HWHM and r correspond to predicted values for an Airy disk imaged at 5500 Å. Note that an approximation as a Gaussian profile is a poor representation of the actual PSF but the determined EHWHM for a single measurement is not far off. From Howell & Merline (1991).

one pixel (plus sign at $r = 0.25$) contains much more flux than any of the remaining ones. A standard Gaussian fit to these data (in this case made by IRAF) is seen to provide a complete misrepresentation of the image profile. Imagine the photometric error one would introduce by assumption of this type of profile and use of its shape as an indication of the total counts observed for this star. The effective FWHM (EFWHM) is defined as the apparent PSF width as determinable from a single undersampled image of a star. We see here that the EHWHM is 0.7 pixels, compared with the expected value (at 5500 Å) of 0.55. The digitization effect present at the highest SSI gain setting leads to an uncertainty of ±379 electrons per DN. The above effects combined lead to an overall relative photometric uncertainty of 5–10% and an absolute spectrophotometric uncertainty of 10–30% for SSI data. These are higher than the 2–5% uncertainties quoted for the WFPC2 camera and are directly in proportion to the greater undersampling and higher CCD gain values used in the Galileo SSI.

Further readings concerning the special conditions and circumstances of CCDs when used for space-based observations can be found with a quick search of the websites of the Hubble Space Telescope and other satellite and spacecraft observatories. Access to numerous internal technical, engineering, and calibration reports is given as well as literature articles containing applications of the findings to astrophysical objects.

7.2 Radiation damage in CCDs

With the launch of the Galileo spacecraft and the Hubble Space Telescope, astronomical imagery with CCDs from outer space began. Today Cassini, Deep Impact, Chandra, XMM-Newton, and a number of other satellitees and space missions (such as the proposed Constellation-X, DUO, ROSITA, and GAIA space missions) have CCD imagers on-board. With these exciting new windows on the Universe come many unexpected effects in the performance and output noise levels of the CCDs involved. The study of radiation damage in CCDs had occurred in a number of military projects, but the low incident flux and low noise levels needed for astronomy required new laboratory work and the development of techniques to deal with or avoid radiation effects altogether (Cameron *et al.*, 2004; Meidinger *et al.*, 2004a; Meidinger *et al.*, 2004b).

The hostile conditions expected in outer space were not the only radiation source to be concerned about for CCDs. Satellites in low Earth orbit, such as the Hubble Space Telescope, pass through the South Atlantic Anomaly

(SAA) periodically, receiving a healthy dose of high energy protons. The Chandra X-ray observatory's largest factor that reduces observing efficiency is the interruption of observations due to passage through the Earth's radiation belts every 2.6 days. X-ray observations are suspended for $\sim$15 hours and the X-ray imager is purposely defocused to minimize damage from low energy (100–200 keV) protons (DePasquale *et al.*, 2004). Solar satellites, such as CAST, are also prone to harsh radiation environments (Kuster *et al.*, 2004). Deep space missions like Galileo and Cassini have a radioisotope thermal electric generator (RTG) to provide power for the spacecraft as well as a neutron dose that bathes the on-board CCD imager. These inherent radiation environments, along with the general space background of cosmic rays and high energy particles from such events as solar flares or planetary magnetic fields, cause both temporary and permanent damage to a CCD in addition to long-term degradation.

Ironically, as CCDs became better astronomical devices in terms of their low read noise and dark currents, they also became much more susceptible to damage by high energy radiation. The SAA, for example, provides about 2000 protons per square centimeter per second with energy of 50–100 MeV, for each passage. Galileo's RTG produced 10^{10} neutrons per square centimeter at the location of the CCD over the expected six-year mission lifetime. Passage through Jupiter's radiation belts near the moon Io was predicted to provide a 2500 rad dose of radiation to the CCD with each orbit. These levels of radiation do indeed cause damage to the CCD involved and methods of monitoring the changes that occur with time and the development of new manufacturing techniques aimed at radiation hardness were needed (McGrath, 1981).

The two major areas of concern in radiation damage to CCDs are (1) high energy photon interactions, which result in fast electrons, which in turn cause simple, localized damage defects and the generation of numerous electron–hole pairs, and (2) nuclear reactions caused by uncharged neutrons or high energy protons, which cause large area defects and are more likely to lead to partial or complete failure of a device (Janesick, Elliott, & Pool, 1988; Janesick, 2001). The first of these radiation induced concerns is called an ionization effect and involves gamma rays or charged particles. The second, involving massive particles, is termed a bulk effect or displacement damage owing to its ability to displace silicon atoms from their lattice positions within the CCD.

Displacement damage can involve single silicon atoms or bulk damage involving clusters of atoms, all removed from their original lattice locations within the CCD. The vacancies remaining in the lattice structure create trapping locations, which in turn cause degraded or no CTE performance for one or more pixels in the array. As the result of lattice stresses, the trap locations become populated by one or more of the doping elements such as phosphorus. The presence of a phosphorus atom within the silicon lattice modifies the band gap energies locally and is thought to be the cause of observed reduced CTE effects (Srour, Hartmann, & Kitazaki, 1986). The CTI performance of the front illuminated CCDs aboard HST suffered radiation damage from exposure to soft protons when passing through the SAA. The damage increased the CTI by more than two orders of magnitude (Grant *et al.*, 2004) and the observatory team has developed a model of the damage to help mitigate its effect on observations.

Repair of some percentage of single lattice displacement defects (i.e., hot pixels) has been accomplished by cycling the CCD to room temperature or higher and back again to operating temperature, a process called annealing. The back-side, thinned SITe CCDs in the HST Advanced Camera for Surveys (ACS) undergo a routine monthly annealing process. Hot pixels (pixels with enhanced dark current of 0.04 electrons/pixel/s or more), appear at a rate of ∼1230 per day in the ACS CCDs. Annealing the detectors will fix about 60–80% of new hot pixels (new since the last anneal) but very few of the older hot pixels are repaired. Figure 7.4 illustrates this procedure for the ACS Wide Field Camera (WFC) CCDs in the ACS.

Bulk defects in CCDs are essentially impossible to repair. It has been noticed, however, that at low temperatures ($< -100°$ C), the trapped charge

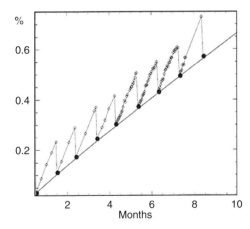

Fig. 7.4. This figure shows the growth in the number of ACS/WFC hot pixels since installation aboard the HST. One can see the lowering of the hot pixel count through monthly anneals as well as the continued overall evolution to increasing numbers. From Clampin *et al.*, 2002.

accumulated at the defect location remains trapped and has little effect on the overall CTE. This temperature dependence has been shown to be proportional to $\exp(-E_T/kT)$, where E_T is the activation energy of the lattice traps (Janesick, Elliott, & Pool, 1988). Thus one way to avoid lattice defects is to operate the CCD at temperatures as low as possible. Interestingly, high temperature operation $(>30°\,C)$ allows trapped charge to be released very quickly, eliminating a deferred charge tail and providing good CTE. These various techniques involving temperature manipulation of a CCD system are often hard to employ practically in space. Additionally, the temporal behavior of the CCD involved can be unpredictable and may be different for each CCD, even those of the same type.

Ionization effects, caused by gamma rays or charged particles, cause a charge buildup in the CCD gate structures and can produce increases in the CCD dark current. Whereas a 150 keV electron is needed to cause an actual silicon atom displacement, only a few eV of energy deposited in the gate insulator is enough to change the potential and cause charge trapping. Even an intense UV flood (Ditsler, 1990; Schaeffer *et al.*, 1990) with 2500 Å photons can cause dark current increases or even render the CCD inoperable. The charge buildup causes new states to exist within the band gap of the silicon leading to easier generation of thermal electrons and thereby an increased dark current. The affected CCD pixels, that is, those that have been damaged by the ionizing radiation, will show increased dark current, while their neighbors will not. Histograms of the amount of dark current produced as a function of pixel signal level often show "spikes" of dark current at specific signal levels. This is taken to indicate a sort of quantized structure in the amount of damage that occurs per radiation site (Janesick, Elliott, & Pool, 1988).

Additional damage to CCDs in space by micrometeoroids has recently been studied (Meidinger *et al.*, 2003) for the CCDs on the XMM-Newton X-ray satellite. Other background increasing radiation events occur as well (Freyberg *et al.*, 2004; Katayama *et al.*, 2004) even if their cause remains a mystery.

Methods of protecting CCDs from radiation effects are varied. The type and amount of radiation expected and the scientific goals of the imager must be carefully weighed to produce the final compromise. For example, the CCD flown on the Galileo mission was initially tested for its ability to withstand gamma radiation similar to that expected in the Jovian magnetic fields, a test that it passed well. It was probably by shear luck that a test or two was also performed to understand its performance when exposed to neutrons. Neutron bombardment revealed an increased dark current was prevalent in the CCDs and a redesign of the imager was needed. To mitigate the problem, the CCD imager had its operating temperature changed to $-130°$ C compared with the original specified value of $-40°$ C. In the case of the Hubble Space Telescope CCDs, increased dark current is not a large factor because of their colder operating temperature, but long-term degrading CTE and QE effects have been seen (Holtzman, 1990; Clampin *et al.*, 2002) and attributed to in-orbit radiation damage. A detailed report of the detectors in the Hubble telescope is contained in Brown (1993) and considerations on improving the ability of CCDs to counteract the effects of radiation are discussed in IEEE Nuclear Science Symposium (1988), Bely, Burrows, & Illingworth (1989), and Janesick (2001).

7.3 CCDs in the UV and EUV (300–3000 Å) spectral range

Progress in the use of CCDs for UV and EUV observations has occurred on two main fronts. Coatings applied to the CCDs to down-convert high energy photons to visible light photons is one method. The other method involves new manufacturing techniques that allow short wavelength photons to penetrate the silicon even given the very small absorption depths at these wavelengths.

Coating the CCD with a UV phosphor has been discussed previously in this book (Chapter 2) with regard to enhancement of the near-UV wavelength range. These same coatings can often increase the QE of a CCD to usable levels, down to wavelengths as short as 500 Å. Lumogen, a UV phosphor that absorbs photons of $\lambda < 4200$ Å and reemits their energy near 5200 Å, is a popular choice. This inexpensive material is thermally deposited onto the CCD surface while under vacuum into a layer about 6000 Å thick. Use of lumogen as a coating delivers a QE from 500–4000 Å of around 15% (yielding about 3% with a solar blind filter in place) and actually increases the intrinsic QE from 5000 to 8000 Å, as it acts as an antireflection coating. Phosphor coatings can be deposited onto either front- or back-side illuminated CCDs (McLean, 1997c; Lesser, 2002). Other coatings such as coronene and Metachrome II are also used for UV and EUV enhancement (Geary *et al.*, 1990; Schempp, 1990; Trauger, 1990).

Modern manufacturing processes have again come to the rescue through the development of techniques that allow the CCD itself to have more sensitivity to short wavelength photons. We have mentioned before that the gate structures of a CCD can absorb short wavelength photons before they enter the actual pixel itself, thus reducing or eliminating the collection probability at these wavelengths. Solutions to this problem consisted of using a back-side illuminated, thinned device or making CCDs with transparent gates. Both of these techniques can be further improved and employed for detection of UV and shorter wavelength photons.

During the thinning process, the back-side of a CCD forms an oxide layer, which can include a surface layer of incomplete bonds that are positively charged. With the very short absorption depths for UV and shorter wavelength photons, any photoelectrons produced within the silicon are more likely to be attracted to the positive bonds where recombination occurs, and not to the pixel potential well for collection, storage, and output. Various techniques have been developed to reduce the number of positive incomplete bond sites that exist in a thinned CCD (Janesick *et al.*, 1985; Janesick *et al.*, 1989; Baily *et al.*, 1990). Additionally, methods that allow removal of the oxide layer produced in the thinning process have been developed and consist of precision etching of the oxide layer under controlled conditions (Bonanno, 1995).

Recent detailed work, aimed at understanding the characteristics and performance of CCDs in the wavelength range of 300–4000 Å, has been performed. The laboratory setup used and various types of short wavelength enhanced CCDs produced and tested are described in Bonanno (1995), which concludes that QE values of 10–60% can be achieved over the wavelength range of 300–2500 Å. These same CCDs also have up to 80% QE at 6000 Å. Phosphor coatings and manufacturing improvements are about equal in their ability to enhance UV and EUV performance; however, both types of improvement appear to show a decrease in their QE with time, once the CCD is cooled and put under vacuum. Contamination by outgassing within the vacuum and subsequent freezing of the contaminants onto the CCD surface are thought to be the most likely causes of the reduced QE.

To overcome the low QE of CCDs at short wavelengths, even after the above enhancements have been performed, some high-energy applications employ standard unmodified CCDs as detectors, preceded in the optical path by a device such as a microchannel plate (MCP). MCPs can produce up to about 500 000 electrons per incident high energy photon, and this electron cloud strikes a phosphor coated photocathode producing visible light photons that are collected and imaged by the CCD (Eccles, Sim, & Tritton, 1983). MCPs operate at high voltages (a few keV) and are inherently solar blind as they require high energy photons for activation. Final QE values of up to 20% are possible with a well-constructed device. This increased QE is the largest advantage of instruments that use intensified CCDs, while poor spatial resolution, phosphor decay effects, and smaller dynamic range (compared with a normal CCD) are the major disadvantages (McLean, 1997c; Longair, 1997).

7.4 CCDs in the X-ray (<500 Å) spectral range

Figure 7.1 provided hints that CCDs may also be useful detectors for the X-ray region of the spectrum, as the absorption depth within silicon rises shortward of about 1000 Å. Figure 7.5 shows us a similar result, only this time we express it as the quantum efficiency of the CCD as a function of photon energy or wavelength. We note that within the X-ray region, back-side thinned CCDs are extremely efficient detectors, approaching a quantum efficiency of 100% at times. The X-ray telescopes aboard XMM-Newton and Chandra use CCDs as their detectors (Longair, 1997; Marshall *et al.*, 2004; Sembay *et al.*, 2004).

X-ray detection by CCDs works in a slightly different manner from detection of optical photons. An incident optical photon creates a photoelectron within the silicon lattice, which moves from the valance to the conduction band and is then held there (in a pixel) by an applied potential. The absorption of an X-ray photon by silicon ejects a free, fast moving, photoelectron of energy $E - b$, where $E = h\nu$ and b is the binding energy of the electron to the silicon atom, typically 1780 eV. As this highly energetic electron moves through the silicon lattice, it produces a trail of electron–hole (e–h) pairs,

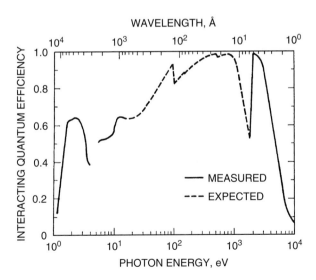

Fig. 7.5. Quantum efficiency for a typical thinned, back-side illuminated CCD from the X-ray to the optical spectral regions. From Janesick *et al.* (1988).

with each requiring an average of 3.65 eV of energy to be produced.[1] Each incident X-ray photon collected produces a measureable number of e–h pairs, thus yielding a method by which one can backtrack and obtain the incident photon energy (Longair, 1997). If all the energy of the free electron went into the e–h pair production, the energy of the incident X-ray could be precisely knowable simply by counting the ADUs produced within the CCD pixels. This property leads to an interesting aspect of X-ray imaging in that one can use the number of photoelectrons produced by an incoming photon to tell its incident energy (wavelength), thereby performing imaging and (crude) X-ray spectroscopy simultaneously.

However, a small undetermined amount of the free electron's energy goes into various phononic states of the silicon lattice, thereby causing some uncertainty in the value of the incident photon's energy. The level of this uncertainty, the "Fano" factor,[2] is so small that to obtain Fano-noise limited CCD performance, the CCD read noise must be less than about 2 electrons (Janesick *et al.*, 1988; Janesick, 2001). Imaging an ^{55}Fe source (used to measure CTE – Chapter 3), would produce a single spectral line at the 5.9 keV Fe Kα energy level while imaging a real astronomical source would produce a crude X-ray spectrum covering the energy range of the detected photons. This type of X-ray spectroscopy was used to produce very low resolution spectra using the imaging capabilities of the ROSAT X-ray satellite.

The Chandra telescope obtains X-ray images and spectroscopy but, in this case, the spectra are not produced by unfolding the images via monitoring image energy deposition, but through the use of gratings to disperse the X-rays (Marshall *et al.*, 2004) in the same manner as discussed for optical spectroscopy in Chapter 6. The imaging and spectroscopy on Chandra both use MIT/LL CCD detectors. These devices are 1024 × 1024 frame transfer CCDs with 24 micron pixels. The frame transfer nature of the CCDs provides fast readout capabilities and therefore can act as an electronic shutter for X-ray observations. Some of the CCDs are front-side illuminated CCDs but these have suffered a lot of damage from the X-rays incident on their (front-side) gate structures (Grant *et al.*, 2004). The back-side illuminated CCDs fare better in terms of radiation damage as well as having overall better QE at lower energy (Figure 7.6).

Figure 7.7 shows an X-ray spectrum of the star Capella obtained with the Chandra observatory using the high energy transmission grating (HETG).

---

[1] Note this value is about equal to the energy of a typical optical photon, which we already know produces one photoelectron.

[2] The term Fano factor is due to U. Fano who, in 1947, formulated a description of the uncertainty in the energy of ion pairs produced in a gas by ionizing radiation.

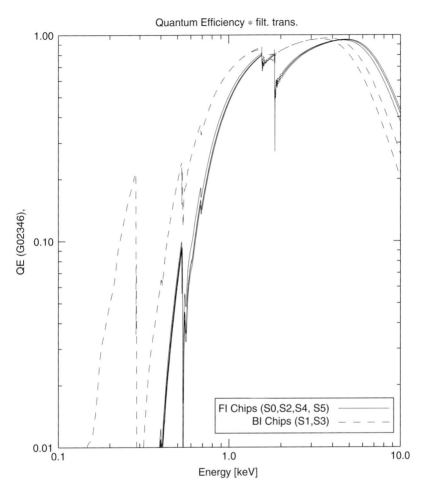

Fig. 7.6. X-ray QE for the CCDs aboard the Chandra X-ray observatory. These QE curves are those of the CCDs convolved with the X-ray filters used. The QE jumps or "edges" seen are caused by inner electronic shell energies of the elements, such as C, used in the X-ray filters.

The spectrum covers the wavelength range from 6–18 Å and shows emission lines (identified in the figure) due to the hot (1 million kelvins or more) stellar corona. XMM-Newton, another X-ray satellite currently in operation, also uses CCDs as detectors. Both of these orbiting X-ray observatories have detailed web pages discussing their telescopes and detectors. A full description is beyond the scope of this volume, but Appendix A contains a number of interesting links to explore.

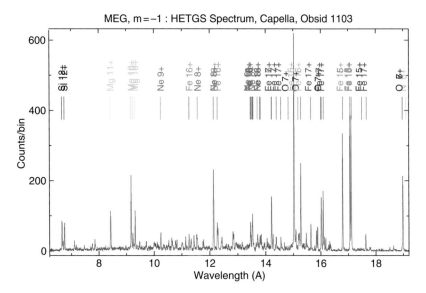

Fig. 7.7. X-ray spectrum of the star Capella obtained with CCDs aboard the Chandra X-ray observatory. The strong emission lines are produced in the hot corona of the star.

7.5 Exercises

1. Make a list of astronomical sources that have their largest flux output at each of the following wavelengths: X-ray, UV, optical, IR, and radio.

2. Which wavelengths require space-based observations?

3. Why do cosmic rays in ground-based CCD images generally produce single pixel events while those in space-based CCD images can produce long streaks and wiggles, as well as single pixel events?

4. Which type of radiation damage has a greater effect on photometric observations from space? Which type affects spectroscopic observations more?

5. Using the idea of charge diffusion and applying it to silicon atoms in a CCD lattice, can you provide a quantitative description of how annealing a CCD "fixes" the majority of hot pixels?

6. Produce a list of items that are required to be different for a CCD camera operating in Earth orbit compared with a similar camera used at a ground-based telescope. Pay particular attention to the plate scale and the A/D process.

7. Describe an observational program that would allow photometric calibration of a CCD camera aboard an interplanetary spacecraft. What level of accuracy would you expect to achieve?

8. Using the locations of the edges in the QE plot for the CCDs (Fig. 7.6), determine which elements and which K shell electron transition is responsible for each.

9. For a CCD used at X-ray wavelengths, describe in detail the procedure that would allow an X-ray image to produce a low resolution spectrum of the imaged source. Convert the x-axis of the X-ray spectrum shown in Figure 7.7 into keV. Make a plot of the relationship between wavelength and number of generated photoelectrons produced for a given X-ray photon in the spectrum.

A

CCD reading list

This appendix provides a reading list covering the aspects of CCD development, research, and astronomical usage. There are so many articles, books, and journal papers covering the innumerable aspects of information on CCDs that the material presented in a book this size or any size can only cover a small fraction of the details of such work. Even the list presented here does not cover all aspects of interest concerning the use of CCDs in astronomy, but it does provide a very good starting point. The growth of information on CCDs has risen sharply over the past ten years and will, no doubt, continue to do so. Thus the student of CCD science must constantly try to keep up with the latest developments both in astronomy and within the field of opto-electronics, both areas where progesss is being made. The internet is a powerful tool to help in this pursuit. Using a good search engine (e.g. Google) type in items such as "deep depletion," or "L3CCD," or "MIT/LL" and you'll get back many items of interest.

Much of the information on CCDs is contained in books devoted to the subject. Numerous SPIE, IEEE, and other conferences publish their proceedings in books as well. Detailed information is available in the scientific literature some of which we reference in this volume. Many refereed articles of interest are not listed here as they are easily searched for via web-based interfaces such as the Astrophysics Data System (ADS).

Numerous non-technical articles related to CCDs have appeared in such magazines as *Astronomy, Sky & Telescope, Scientific American*, and *CCD Astronomy*.[1] The reader is cautioned that in some of these descriptions of CCDs, the details are not always completely correct nor have the methodologies always been rigorously tested. However, some of the "popular" articles contain a wealth of information that would be unlikely to appear elsewhere.

---

[1] *CCD Astronomy* is no longer published but has been incorporated into *Sky & Telescope*.

We refer the reader to the websites of these magazines for a list of relevant articles.

Every major observatory is a storehouse of knowledge on CCDs. They have websites filled with useful information, large numbers of engineering and technical reports covering aspects of instrument design and construction, manuals for the observatory instruments available, and often newsletters that contain very worthwhile information on their particular CCDs. Many laboratories and companies working on CCDs also have websites that are extremely valuable (see Appendix B).

A.1 General CCD references

Listed below are some general references containing details on CCDs. They are listed by year of publication and the end of the list contains periodic references.

Eccles, M. J., Sim, M. E., & Tritton, K. P., 1983, *Low Light Level Detectors in Astronomy*, Cambridge University Press.

Borucki, W. & Young, A. (eds.), 1984, *Proceedings of the Workshop on Improvements to Photometry*, NASA Conference Publication 2350.

Dereniak, E. & Crowe, D., 1984, *Optical Radiation Detectors*, Wiley.

Hearnshaw, J. B. & Cottrell, P. L. (eds.), 1986, *Instrumentation and Research Programmes for Small Telescopes*, D. Reidel.

Walker, G., 1987, *Astronomical Observations*, Cambridge University Press.

Borucki, W. (ed.), 1988, *Second Workshop on Improvements to Photometry*, NASA Conference Publication 10015.

Robinson, L. B. (ed.), 1988, *Instrumentation for Ground-Based Optical Astronomy*, Springer-Verlag.

Jacoby, G. H. (ed.), 1990, *CCDs in Astronomy*, ASP Conference Series Vol. 8.

Philip, A. G. D., Hayes, D., & Adelman, A. (eds.), 1990, *CCDs in Astronomy II: New Methods and Applications of CCD Technology*, L. Davis Press.

Wall, J. V. & Boksenberg, A. (eds.), 1990, *Modern Technology and Its Influence on Astronomy*, Cambridge University Press.

Berry, R., 1991, *Introduction to Astronomical Image Processing: A Comprehensive Guide to CCD Image Enhancement for the IBM-PC and Compatibles*.

Blouke, M., 1991, *Charge-Coupled Devices and Solid State Optical Sensors II*, SPIE, Vol. 1447.

Buil, C., 1991, *CCD Astronomy: Construction and Use of an Astronomical CCD Camera.*

Berry, R., 1992, *Choosing and Using a CCD Camera: A Practical Guide to Getting Maximum Performance from Your CCD.*

Howell, S. B. (ed.), 1992, *Astronomical CCD Observing and Reduction Techniques*, ASP Conference Series Vol. 23.

Butler, C. J. & Elliot, I. (eds.), 1993, *Stellar Photometry – Current Techniques and Future Developments*, IAU Colloquium 136, Cambridge University Press.

Kilkeny, D., Lastovica, E., & Menzies, J. W. (eds.), 1993, *Precision Photometry*, South African Astronomical Observatory Publication.

Berry, R., 1994, *The CCD Camera Cookbook: How to Build Your Own CCD Camera.*

Sze, S. M., 1994, *Semiconductor Sensors*, Wiley.

Philip, A. G. D., Janes, K., & Upgren, A. (eds.), 1995, *New Developments in Array Technology and Applications*, IAU Symposium No. 167, Kluwer.

Ratledge, D., 1996, *The Art and Science of CCD Astronomy (Practical Astronomy).*

Reike, G., 1996, *The Detection of Light from Ultraviolet to the Submillimeter*, Cambridge University Press.

Lena, P., 1997, *A Practical Guide to CCD Astronomy (Practical Astronomy Handbooks).*

Longair, M., 1997, *High Energy Astrophysics*, Vols. 1 & 2, Cambridge University Press.

McLean, I. S., 1997, *Electronic Imaging in Astronomy*, Wiley.

Beletic, J. & Amico, P., 1998, *Optical Detectors for Astronomy: Proceedings of an ESO CCD Workshop held in Garching, Germany, October 8–10, 1996*, Kluwer Academic Press.

Holst, G. C., 1998, *CCD Arrays, Cameras, and Displays*, SPIE Press.

Martinez, P. & Klotz, A., 1998, *A Practical Guide to CCD Astronomy*, Cambridge University Press.

Janesick, J., 2001, *Scientific Charge-Coupled Devices*, SPIE press, PM83.

Wodaski, R., 2002, *The New CCD Astronomy: How to Capture the Stars With a CCD Camera in Your Own Backyard.*

Kang, M. G., 2003, *CCD and CMOS Imagers*, SPIE Milestone Series, MS 177.

Amico, P. *et al.*, 2004, *Scientific Detectors for Astronomy*, Kluwer Academic Press.

Nathan, A. & Li, F., 2004, *CCD Image Sensors in Deep-Ultraviolet: Degradation Behavior and Damage Mechanisms.*

Periodicals

Photonics Handbook. Yearly editions available from Laurin Publishing Company Inc., Pittsfield, MA.

Experimental Astronomy. This journal is a wealth of information on CCDs as well as imagers used at other wavelengths, astronomical instrumentation, and current research areas in astronomical detectors.

Optical Engineering. The primary journal of the International Society for Optical Engineers, has numerous special editions devoted to CCDs.

Proceedings of SPIE. The International Society for Optical Engineering. Numerous volumes over the past twenty years or so. Of special interest are the volumes on "Instrumentation in Astronomy," and "Advanced Technology Optical Telescopes."

B

CCD manufacturers: websites & information

Table B.1 lists the websites for current CCD manufacturers and government laboratories engaged in CCD development (as of 2005).

Table B.1. *CCD and CCD Camera Manufacturers*

Company Name	Web Address
E2V (Marconi, EEV)	http://e2vtechnologies.com/
Dalsa	http://www.dalsa.com/
STA	http://www.sta-inc.net/recentnews.html
Kodak	http://www.kodak.com/global/en/digital/ ccd/sensorsMain.jhtml
Orbit	http://www.orbitsemi.com/
Philips	http://www.semiconductors.philips.com/
Sarnoff	http://www.sarnoff.com/
SITe / Pixelvision	http://www.site-inc.com/
TI	http://www.ti.com/
Thomson CSF	http://www.thomson.net
LLNL	http://www-phys.llnl.gov/
LBL	http://www-ccd.lbl.gov/
MIT/LL	http://www.ccd.ucolick.org/lincoln/ lincoln.html
EG&G Reticon (PerkinElmer Optoelectronics)	http://www.perkinelmer.com/
Loral (Fairchild)	http://www.fairchildimaging.com/main/
SBIG	http://www.sbig.com/
Apogee	http://www.ccd.com/
Andor	http://www.andor-tech.com/

C

Some basics of image displays and color images

Most computer screens and image displays in use are 8-bit devices. This means that the displays can represent data projected on them with $2^8 = 256$ different greyscale levels or data values of resolution. These greyscale levels can represent numeric values from 0 to 255 and it is common to only have about 200 levels actually available to the image display for representing data values with the remaining 50 or so values reserved for graphical overlays, annotation, etc. If displaying in color (actually pseudo-color), then one has available about 200 separate colors, each with a possible grey value of 0–255, or the famous "16 million possible colors" listed in many computer ads (see below).

On the display, the color black is represented by a value of zero (or in color by a value of zero for each of the three color guns, red (R), green (G), and blue (B)). White has $R = G = B = 255$, and various grey levels are produced by a combination of $R = G = B = N$, where N is a value from 0 to 255. Colors are made by having $R \neq G \neq B$ or any combination thereof in which all three color guns are not operated at the same intensity. A pure color, say blue, is made with $R = G = 0$ and $B = 255$ and so on. You may have noticed that color printers have three (or four) colors of ink in them. They contain cyan, blue, and magenta (and black) inks, which are used in combination to form all the output colors. This difference (cyan etc. vs. RGB) in the choice of colors is simply because display screens mix light whereas printers mix ink to form specific colors.

Terms one hears but rarely uses in astronomy are hue, saturation, and brightness. Hue means the color of the image, saturation is the relative strength of a certain color (fully saturated $= 1$), and brightness is the total intensity of a color where black $= 0$. When you change colors (RGB) you are really changing the hue, saturation, and brightness of the image display. These three terms are fully explored in Gonzalez & Woods (1993) as well as in almost any text introducing image processing techniques.

Almost all CCD data obtained today have a dynamic range of much greater then 8 bits. Thus, in order to display the CCD image, some form of scaling must be performed to allow the image to be shown on a display with only 8 bits. A common technique (often performed by the software without user intervention) is called linear scaling. This type of scaling divides the entire true data range into say 200 equal bins, where each bin of data is represented by 1 of the 0–200 available greyscale levels. For example, if an image has real data values in the range from 0 to 100 000 ADUs, linear scaling will place the real data values between 0 and 500 ADU into the first scaled bin and will display them as a 0 on the screen. If your image is such that all the interesting astronomical information has real values of 0 to 2000, this linear scaling scheme will represent all the real image information for the values of 0 to 2000 within only 4 display bins, those having values of 0–4.

To avoid such poor scaling and loss of visual information, two alternatives generally exist: one uses a linear scaling but within a specific data window and one uses a different type of scaling altogether. The first option is accomplished by having the software again perform a linear scaling but this time using its 200 output display levels to scale image data values only within the data window of say 0 to 2000. Different scaling options allow for nonlinear modes such as log scaling, exponential scaling, histogram equalization, and many others. These are easily explored in any of the numerous image processing software packages used today.

A method commonly used to aid the eye when viewing a displayed image is that of interactive greyscale manipulation. You probably know this as changing the image stretch or contrast and perform it via movement of a mouse or trackball while displaying an image. The actual change that is occurring is a modification of the relation between the input data values (the 8-bits loaded into memory as chosen for display) and those output to the display screen. The software mechanism that controls this is called a look-up table or LUT. Some sample LUTs are shown in Figure C.1, where we see the relation of input to output data value. For greyscale images, all three color LUTs are moved in parallel, while in pseudo-color mode (see below), each color LUT can be individually controlled. Changes in the slope and intercept of the LUT control changes in the image brightness, contrast, and color. Terms such as "linear stretch" simply refer to a LUT using a linear transformation function.

Further information on general image processing techniques can be found in Gonzalez & Woods (1993), while additional details on image displays can

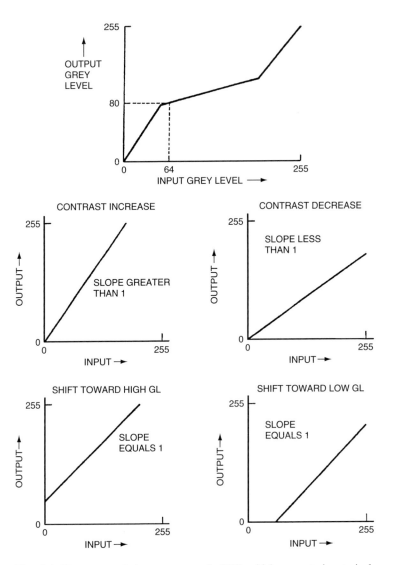

Fig. C.1. The top panel shows an example LUT, which converts input pixel values or grey levels to output grey levels on the video display. In this particular case, an input pixel value of 64 will be displayed with a grey level value of 80. Note that the relationship between input and output is not a single-valued linear transformation. The bottom panel shows various linear stretch operations. If the mapping passes through the origin and has a slope >1, the effect will be to increase the output contrast. A slope of <1 decreases the contrast. If the linear mapping intersects the *y* axis (i.e., the output value axis) the input values will be systematically shifted to higher output grey levels. Intersecting the horizontal axis causes a net shift to lower output grey levels. From Gonzalez & Woods (1993).

be found in Hanisch (1992). User manuals for astronomical image processing software packages such as IRAF and MIDAS often have numerous examples and routines useful for dealing with CCD imagery.

C.1 False color and pseudo-color images

Observations with Charge-Coupled Devices (CCDs) do not really generate images, they produce a 3-D set of numbers: x, y, and intensity (z). We have just seen how these numbers are displayed as an image and how the use of a look-up table allows the translation from pixel value (z) to displayed grey scale. Sometimes, we wish to present our results in color either for dramatic effect or for science productivity, or at times, a little of both. Additionally, color images often truly help us to see the details better.

A black-and-white image may make the digital data more understandable, but the number of different grey tones that the human eye can separate is very limited. Our eyes can only see or interpret 20–30 grey steps yet an 8-bit display shows us 200 or more steps on a contrast scale. On the other hand, our eyes can separate 20 000 or more different color tints or shades making pseudo and false color images much better at showing the real variation in the original data.

Pseudo-color images are single displayed images (usually 8-bits) in which each grey level is assigned a color. A simply approach may be to assign a LUT that is a ramp of red values from a light red to a dark red, with an ever increasing redness as the LUT goes form 0 to 255. A more complex pseudo-color representation may be to take three colors, red, green, and blue, and assign LUT values as follows: 0–85 is a red ramp, 86–171 is a green ramp, and 172–255 is a blue ramp. Another scheme might use these same three LUT ranges but only one color. In this mode, the real intensity values are represented by color that wraps around at each numeric boundary. The pixel values of 9, 95, and 181 will be displayed as the same color. In the end, we can use the 8-bit display to show a color image in which the grey levels of 0–255 are assigned a single color per datum value. There may be logic to the color assignment such as showing an image of the planet Neptune in shades of blue to express coldness or showing the crab nebula in red to infer hotness. This type of color scheme tries to help us interpret the scientific meaning of the image through casual visual inspection. Another choice of pseudo-color may be to try to separate image features close in value through use of color. Let's say we have a planetary nebula image and we wish to separate the central star from the gaseous nebula for contrast in a presentation. However, the central star is faint and its pixel values are 100–200 ADU while the nebula itself is 400–600 ADU. Using the above scheme of color ramps would assign shades of the same color to both components, reducing their contrast in the displayed image. However, use of a different color for each block of say 100 ADU levels will allow the user to stretch the 8-bit displayed image such that the star and the nebula are separated by a color boundary, thereby making them distinct in the final display.

A "false color" image does not mean that the data are wrong or that the picture is deceiving you. The term simply means that the image is not a real color photograph. False color images are composed of three separate 8-bit images displayed simultaneously (i.e., added together) allowing 24 bits or roughly 16 million possible color values to be represented. Each 8-bit color plane has a unique LUT of color values for its 256 possible numeric levels. An example of a display scheme may be to assign each 8-bit plane to a ramp of red, green, and blue. Colors are produced when the three LUTs overlap (the three images are displayed simultaneously) for any given pixel. If the red, green, blue (RGB) values are set to (255,255,255) the color appears white, (0,0,0) is black, (255,0,0) is pure red, (255,0,255) is pure purple, etc. The primary colors (RGB), when mixed in groups of two, produce the secondary colors of cyan, magenta, and yellow. The production of color and shades of color relies on the combination of the three LUTs as laid out in a chromaticity diagram or color wheel (Rector *et al.*, 2005). This method of producing various colors works in much the same way as a color television. If you look closely at a color TV screen, you'll see that it is made up of many sets of three circular or hexagonal color regions (red, green, blue) and by varying the intensity in each, the color is changed. Your eye can not resolve (from far away) the three segments and thus blends them together to form the desired color. Color prints follow this scheme as well, usually printing images with three passes, one for each of three printed colors.

Let us look at an example. We wish to present a color image of a star-forming region in Orion using three digital images. One of the images was obtained with a CCD camera through an [O III] filter centered on the 4959 Å and 5007 Å emission lines. The second image was obtained with the same camera but using an Hα filter centered at 6563 Å. The third image, produced by the same camera, used a broad-band U band filter centered at 3500 Å. One choice of color presentation is to make each image a pseudo-color image assigning each a color LUT in red, green, and blue. The three images can then be displayed beside each other or printed out and viewed next to each other.

Another method is to produce a false color image by combining the three 8-bit images into a single 24-bit image. We might assign each image a color scheme in a similar way as just described (or not) and then combine the three into a single image. Tweaking of the color LUTs can then be performed to produce the final scientifically useful, as well as eye-pleasing, image. Other wavelengths, such as X-ray or infrared light images, can also be combined with or without an optical component to produce false color images. In these images, the color choice is often a matter of personal taste, and is used in a manner generally associated with the intensity or brightness of the radiation from different regions of the image.

For example, in a greyscale or black and white Chandra X-ray image of a supernova remnant, the darker shades of grey might represent the most intense X-ray emission, the lighter shades of gray could represent the areas of less intense emission, and the white areas could be used to show areas of little to no emission. In a color version, lighter colors such as yellow or orange could represent areas of high X-ray intensity, orange to red areas of lower intensity, and black representing little or no emission. This false color image representation provides the eye with a seemingly hot region (bright colors) for the highest X-ray emission and ends with darker, low emission regions in black. The color assignments in each image LUT in this case follow variations in intensity of the X-ray counts, which in reality are associated with variations in the density, or concentration, of hot gas. A typical method used to produce false color images from Chandra data is to use three images obtained in the energy bands of 0.3 to 1.55 keV, 1.55 to 3.34 keV, and 3.34 to 10 keV, respectively. The intensity value within each image is mapped to a color LUT value of 0–255 and the three images are then combined to produce a false color X-ray image.

If we combine three optical images in a manner akin to how our eyes would see it, e.g., with similar intensity ratios for red, green, and blue light, the image is called a "true color" image. As usual, the best way to learn all about color imagery is to experiment yourself. Programs such as "The Gimp," XV, Photoshop, and others can be used for this purpose. An excellent account of how to produce color astronomical images including all the factors and intricacies involved, as well as examples, is provided in Rector *et al.* (2005).

For additional information about color imagery, explore these excellent websites:

http://observe.ivv.nasa.gov/nasa/exhibits/learning/learning_0.html
http://observe.ivv.nasa.gov/nasa/education/reference/main.html
http://hawaii.ivv.nasa.gov/space/hawaii/vfts/oahu/rem_sens_ex/rsex.
spectral.1.html

http://www.photo.net/photo/edscott/ms000050.html
http://www.allthesky.com/articles/imagecolor.html
http://chandra.harvard.edu/photo/false_color.html

C.2 Exercises

C.1. Discuss the differences between pseudo-color and true color.

C.2. How does the process of histogram equalization really work? Why might it be useful to use when viewing an image on a display device? Display various CCD images and apply histogram equalization to them. Does the process help in the way that you thought it would?

C.3. Display an image that has more than 8 bits of information per pixel. (Most present-day CCD images are 16–32 bits.) Now display the lowest 8 bits and the highest 8 bits. How does the appearance of the image change? Can you make sense of this change based on your knowledge of the image? How does your display device "decide" which 8 bits to show you by default?

C.4. What are the RGB LUT assignments needed to produce the following colors: yellow, cyan, green, light green, brown, silver?

C.5. Define what is meant by complementary colors. What do the RGB LUT assignments look like for such colors?

C.6. Use one of the programs mentioned in this section (or another of your choice) and use three CCD images to produce three pseudo-color images and a false color image. What are the advantages of the pseudo-color images? How about the false color image?

References

Adams, M., Christian, C., Mould, J., Stryker, L., & Tody, D., 1980, *Stellar Magnitudes from Digital Pictures*, Kitt Peak National Observatory publication.

Alard, C., 2000, *Astron. Astrophys.*, **144**, 363.

Alcock, C., *et al.*, 1999, *Ap. J.*, **521**, 602.

Amelio, G., Tompsett, M., & Smith, G., 1970, *Bell Systems Technical Journal*, **49**, 593.

Anderson, J. & King, I., 2003, *Publ. Astron. Soc. Pac.*, **115**, 113.

Baily, P., *et al.*, 1990, *Proc. SPIE*, **1344**, 356.

Baum, W. A., Thomsen, B., & Kreidl, T. J., 1981, in *Solid State Imagers for Astronomy*, eds. J. C. Geary & D. W. Latham, *Proc. SPIE*, **290**, 24.

Belton, M., *et al.*, 1992, *Space Sci. Rev.*, **60**, 413.

Bely, P.-Y., Burrows, C., & Illingworth, G. (eds.), 1989, *The Next Generation Space Telescope*, Space Telescope Science Institute publication.

Bland, J. & Tully, R., 1989, *Astron. J.*, **98**, 723.

Blouke, M., Yang, F., Heidtmann, D., & Janesick, J., 1988, in *Instrumentation for Ground-Based Optical Astronomy*, ed. L. B. Robinson, Springer-Verlag, p. 462.

Bonanno, G., 1995, in *New Developments in Array Technology and Applications*, eds. A. G. D. Philip, K. A. Janes, & A. R. Upgren, Kluwer, p. 39.

Bonanno, G., 2003, *MnSAI*, **74**, 800.

Born, M. & Wolf, E., 1959, *Principles of Optics*, MacMillan, Chap. VIII.

Boulade, O., Vigroux, L., Charlot, X., *et al.*, 1998, in *Optical Astronomical Instrumentation*, ed. S. D'Odorico, *Proc. SPIE*, **3355**, 614.

Bowen, I. S., 1960a, in *Astronomical Techniques*, ed. W. A. Hiltner, University of Chicago Press, Chap. 2.

Bowen, I. S., 1960b, in *Telescopes*, eds. G. Kuiper & B. Middlehurst, University of Chicago Press, p. 1.

Boyle, W. & Smith, G., 1970, *Bell Systems Technical Journal*, **49**, 587.

Brammertz, G., *et al.*, 2004, *SPIE*, **5499**, 269.

Brar, A., 1984, *AURA Engineering Technical Report No. 76*.

Broadfoot, A. L. & Kendall, K., 1968, *J. Geophys. Res. Space Physics*, **73**, 426.

Brown, R. (ed.), 1993, *The Future of Space Imaging*, Space Telescope Science Institute publication, Chap 8.

Buonanno, R. & Iannicola, G., 1989, *Publ. Astron. Soc. Pac.*, **101**, 294.

Burke, B., *et al.*, 2004, *SPIE*, **5499**, 185.

Burke, B., Reich, R., Savoye, E., & Tonry, J., 1994, *IEEE T. Electron Dev*, **41**, 2482.

Cameron, R., *et al.*, 2004, *SPIE*, **5488**, 572.

Chiu, L.-T., 1977, *Astron. J.*, **82**, 842.

Chromey, F. & Hasselbacher, D. A., 1996, *Publ. Astron. Soc. Pac.*, **108**, 944.

Clampin, *et al.*, 2002, *Exp. Astron.* **14**, 107

Cochran, A., 1995, in *New Developments in Array Technology and Applications*, eds. A. G. D. Philip, K. A. Janes, & A. R. Upgren, Kluwer, p. 251.

Corbally, C., 1995, in *New Developments in Array Technology and Applications*, eds. A. G. D. Philip, K. A. Janes, & A. R. Upgren, Kluwer, p. 241.

Cropper, M., *et al.*, 2003, *Mon. Not. R. Astron. Soc.*, **344**, 33.

Dacosta, G., 1992, in *Astronomical CCD Observing and Reduction Techniques*, ASP Conference Series Vol. 23, ed. S. Howell, p. 90.

Davis, L. E., 1994, *A Reference Guide to the IRAF/DAOPHOT Package*, NOAO publication.

DePasquale, J., *et al.*, 2004, *SPIE*, **5165**, 554.

Delamere, A., Atkinson, M., Rice, J., Blouke, M., & Reed, R., 1990, in *CCDs in Astronomy*, ASP Conference Series Vol. 8, ed. G. H. Jacoby, p. 269.

DePoy, D., *et al.*, 2004, *SPIE*, **5492**, 452.

DeVeny, J., 1990, *The Gold Camera Instrument Manual: A CCD Spectrograph for the 2.1 Meter Telescope*, NOAO Instrument Manual Publication.

Dhillon, V. & Marsh, T., 2001, *New Astron. Rev.*, **45**, 91.

Dhillon, V., Rutten, R., & Jorden, P. R., 1993, *Gemini*, Issue 41, p. 7.

Diego, F., 1985, *Publ. Astron. Soc. Pac.*, **97**, 1209.

Ditsler, W., 1990, in *CCDs in Astronomy*, ASP Conference Series Vol. 8, ed. G. H. Jacoby, p. 126.

Djorgovski, S., 1984, in *Proceedings of the Workshop on Improvements to Photometry*, eds. W. J. Borucki & A. Young, NASA Conf. Publ. 2350.

Eccles, M., Sim, E., & Tritton, K., 1983, *Low Light Level Detectors in Astronomy*, Cambridge University Press.

Everett, M. & Howell, S. B., 2001, *Publ. Astron. Soc. Pac.*, **113**, 1428.

Everett, M., *et al.*, 2002, *Publ. Astron. Soc. Pac.*, **114**, 656.

Faber, S. & Westphal, J. (eds.), 1991, *WFPC Science Verification Report*, Space Telescope Science Institute.

Fehrenbach, C., 1967, in IAU Symposium 30, eds. A. Batten & D. Heard, Academic Press, p. 65.

Florentin-Nielsen, R., Anderson, M., & Nielsen, S., 1995, in *New Developments in Array Technology and Applications*, eds. A. G. D. Philip, K. A. Janes, & A. R. Upgren, Kluwer, p. 207.

Freyberg, M., *et al.*, 2004, *SPIE*, **5165**, 112.

Furenlid, I. & Meylon, T., 1990, in *CCDs in Astronomy II: New Methods and Applications of CCD Technology*, eds. A. G. D. Philip, D. Hayes, & S. Adelman, L. Davis Press, p. 13.

Geary, J., Torres, G., Latham, D., & Wyatt, W., 1990, in *CCDs in Astronomy*, ASP Conference Series Vol. 8, ed. G. H. Jacoby, p. 40.

Gehrels, T., Marsden, B., McMillian, R., & Scotti, J., 1986, *Astron. J.*, **91**, 1242.

Gibson, B., 1991, *J. R. Astron. Soc. Canada*, **85**, 158.

Gibson, B. & Hickson, P., 1992, *Mon. Not. R. Astron. Soc.*, **258**, 543.

Gillespie, B., *et al.*, 1995, ASP Conf. Series Vol. 87, p. 97.

Gilliland, R., 1992, in *Astronomical CCD Observing and Reduction Techniques*, ASP Conference Series Vol. 23, ed. S. Howell, p. 68.

Gilliland, R. L., *et al.*, 1993, *Astron. J.*, **106**, 2441.

Girard, T., *et al.*, 2004, *Astron. J.*, **127**, 3060.

Gonzalez, R. & Woods, R., 1993, *Digital Image Processing*, Addison-Wesley.

Gorjian, V., Wright, E., & Mclean, I., 1997, *Publ. Astron. Soc. Pac.*, **109**, 821.

Grant, C., Bautz, M., Kissel, S., & LaMarr, B., 2004, *SPIE*, **5501**, 177.

Groom, D., 2002, *Exp. Astron.* **14**, 45.

Groom, D., 2004, in *Scientific Detectors for Astronomy*, eds. P. Amico, J. Beletic, & J. Beletic, 2004, Kluwer Academic Publishers, p. 81.

Gudehus, D., 1990, in *CCDs in Astronomy*, ASP Conference Series Vol. 8, ed. G. H. Jacoby, p. 356.

Gullixson, C., 1992, in *Astronomical CCD Observing and Reduction Techniques*, ASP Conference Series Vol. 23, ed. S. Howell, p. 130.

Gunn, J. & Westphal, J., 1981, *Proc. SPIE*, **290**, 16.

Gunn, J., *et al.*, 1998, *Astron. J.*, **116**, 3040.

Hanisch, R., 1992, in *Astronomical CCD Observing and Reduction Techniques*, ed. S. B. Howell, ASP Conf. Series Vol. 23, p. 285.

Hayes, D. S. & Latham, D. W., 1975, *Astrophys. J.*, **197**, 593.

Hendon, A. & Kaitchuck, R., 1982, *Astronomical Photometry*, Van Nostrand Reinhold.

Hickson, P., Borra, E., Cabanac, R., Content, R., Gibson, B., & Walker, G., 1994, *Astrophys. J. Lett.*, **436**, L201.

Hickson, P. & Richardson, E. H., 1998, *Publ. Astron. Soc. Pac.*, **110**, 1081.

Hiltner, W. A. (ed.), 1962, *Astronomical Techniques*, Vol. II of *Stars and Stellar Systems*, University of Chicago Press.

Holland, S., 2004, in *Scientific Detectors for Astronomy*, eds. P. Amico, J. Beletic, & J. Beletic, 2004, Kluwer Academic Publishers, p. 95.

Holland, S., 2002, *Exp. Astron.*, **14**, 83.

Holtzman, J. A., 1990, *Publ. Astron. Soc. Pac.*, **102**, 806.

Holtzman, J., *et al.*, 1995a, *Publ. Astron. Soc. Pac.*, **107**, 1065.

Holtzman, J., *et al.*, 1995b, *Publ. Astron. Soc. Pac.*, **107**, 156.

Honeycutt, R. K., 1992, *Publ. Astron. Soc. Pac.*, **104**, 435.

Horne, K., 1986, *Publ. Astron. Soc. Pac.*, **98**, 609.

Howell, S. B., 1989, *Publ. Astron. Soc. Pac.*, **101**, 616.

Howell, S. B., 1992, in *Astronomical CCD Observing and Reduction Techniques*, ASP Conference Series Vol. 23, ed. S. Howell, p. 105.

Howell, S. B., 1993, in *Stellar Photometry – Current Techniques and Future Developments*, eds. C. J. Butler & I. Elliott, IAU Colloquium 136, Cambridge University Press, p. 318.

Howell, S. B., *et al.*, 2005, *Publ. Astron. Soc. Pac.*, **117**, Nov Issue.

Howell, S. B., Everett, M., Esquerdo, G., David, D., Weidenschilling, S., & Van Lew, T., 1999, in *Precision CCD Photometry*, eds. E. Craine & D. Crawford, ASP Conference Series, Vol. 189.

Howell, S. B., Everett, M., & Ousley, D., 1999, in *Third Workshop on Improvements to Photometry*, eds. W. Borucki & L. Lasher, NASA Conference Publication, NASA/CP-2000-209614, p.79.

Howell, S. B., Everett, M., Tonry, J. L., Pickles, A., & Dain, C., 2003, *Publ. Astron. Soc. Pac.*, **115**, 1340.

Howell, S. B. & Jacoby, G., 1986, *Publ. Astron. Soc. Pac.*, **98**, 802.

Howell, S. B., Koehn, B., Bowell, E. L. G., & Hoffman, M., 1996, *Astron. J.*, **112**, 1302.

Howell, S. B. & Merline, W., 1991, Results from the November 1990 Galileo Satellite SSI Calibration, *Memorandum to Galileo SSI Team*, 8 July 1991 & Addenda A (19 Aug. 1991) and B (5 Oct. 1991), Galileo SSI Team memoranda, Galileo Satellite Project, NASA/JPL.

Howell, S. B., Mitchell, K. J., & Warnock, A., III, 1988, *Astron. J.*, **95**, 247.

IEEE Nuclear Science Symposium, 1988, Radiation Damage in Scientific Charge-Coupled Devices, *IEEE Trans.*, 42.

Jacoby, G., *et al.*, 2002, *SPIE*, **4836**, 217.

Jacoby, G., Hunter, D., & Christian, C., 1984, *Ap. J. Suppl.*, **56**, 257.

Janesick, J., 2001, *Scientific Charge-Coupled Devices*, SPIE Press, Bellingham, WA.

Janesick, J., 2004, in *Scientific Detectors for Astronomy*, eds. P. Amico, J. Beletic, & J. Beletic, 2004, Kluwer Academic Publishers, p. 103.

Janesick, J., *et al.*, 2002, *Exp. Astron.*, **14**, 33.

Janesick, J. & Blouke, M., 1987, *Sky and Telescope Magazine*, September, **74**, 238.

Janesick, J. & Elliott, T., 1992, in *Astronomical CCD Observing and Reduction Techniques*, ASP Conference Series Vol. 23, ed. S. Howell, p. 1.

Janesick, J., Elliott, T., Bredthauer, R., Chandler, C., & Burke, B., 1988, *Optical and Optoelectronic Applied Science and Engineering Symposium; X-ray Instrumentation in Astronomy*, SPIE, Bellingham, WA.

Janesick, J., Elliot, T., Collins, M., Blouke, M., & Freeman, J., 1987a, *Opt. Eng.*, **26**, 692.

Janesick, J., Elliot, T., Collins, S., Daud, T., Campbell, D., & Garmire, G., 1987b, *Opt. Eng.*, **26**, No. 2.

Janesick, J., Elliot, T., Daud, T., McCarthy, J., & Blouke, M., 1985, in *Solid State Imaging Arrays*, eds. K. N. Prettyjohns & E. L. Dereniak, *Proc. SPIE*, **570**, 46.

Janesick, J., Elliot, T., Fraschetti, G., Collins, S., Blouke, M., & Corrie, B., 1989, *Proc. SPIE*, **1071**, 153.

Janesick, J., Elliott, T., & Pool, F., 1988, IEEE Nuclear Science Symposium, Orlando, FL.

Janesick, J., Hynecek, J., & Blouke, M., 1981, *Proc. Soc. Photo-Opt. Instr. Eng.*, *SPIE*, **290**, 165.

Jorden, P. R., Deltorn, J.-M., & Oates, A. P., 1993, *Gemini*, Issue 41, p. 1.

Jorden, P. R., Deltorn, J.-M., & Oates, A. P., 1994, in *Instrumentation for Astronomy VIII, Proc. SPIE*, **2198**, p. 57.

Jorden, P. R., Pool, P., & Tulloch, S., 2004, in *Scientific Detectors for Astronomy*, eds. P. Amico, J. Beletic, & J. Beletic, 2004, Kluwer Academic Publishers, p. 115.

Joyce, R., 1992, in *Astronomical CCD Observing and Reduction Techniques*, ASP Conference Series Vol. 23, ed. S. Howell, p. 258.

Katayama, H., *et al.*, 2004, *A. & A.*, **414**, 767.

Kells, W., *et al.*, 1998, *Publ. Astron. Soc. Pac.*, **110**, 1487.

Kimble, R. A., Goudfrooij, P., & Gilliland, R., 2000, *SPIE*, **4013**, 532.

King, I., 1971, *Publ. Astron. Soc. Pac.*, **83**, 199.

King, I., 1983, *Publ. Astron. Soc. Pac.*, **95**, 163.

Krist, J., 2004, *SPIE*, **5499**, 328.

Kreidl, T. J., 1993, in *Stellar Photometry – Current Techniques and Future Developments*, eds. C. J. Butler & I. Elliott, IAU Colloquium 136, Cambridge University Press, p. 311.

Kuster, M., *et al.*, 2004, *SPIE*, **5500**, 139.

Lasker, B. M., *et al.*, 1990, *Astron. J.*, **99**, 2019.

Leach, R., 1995, in *New Developments in Array Technology and Applications*, eds. A. G. D. Philip, K. A. Janes, & A. R. Upgren, Kluwer, p. 49.

Lesser, M., 1990, in *CCDs in Astronomy*, ASP Conference Series Vol. 8, ed. G. H. Jacoby, p. 65.

Lesser, M., 1994, in *Instrumentation in Astronomy VIII*, eds. D. L. Crawford & E. R. Craine, *Proc. SPIE*, **2198**, 782.

Lesser, M., 2002, *Exp. Astron.*, **14**, 77.

Lesser, M., 2004, in *Scientific Detectors for Astronomy*, eds. P. Amico, J. Beletic, & J. Beletic, 2004, Kluwer Academic Publishers, p. 137.

Longair, M. S., 1997, *High Energy Astrophysics*, Vol. 1, Cambridge University Press.

Lucy, L. B., 1975, *Astron. J.*, **79**, 745.

MacConnell, D., 1995, in *The Future Utilisation of Schmidt Telescopes*, eds. J. Chapman, R. Cannon, S. Harrison, & B. Hidayat, ASP Conference Series Vol. 84, p. 323.

Mack, J., *et al.*, 2002, in *The 2002 HST Calibration Workshop*, p. 23.

Mackay, C., 1986, *Annu. Rev. Astron. Astr.*, **24**, 255.

Marshall, H., Dewey, D., & Ishibashi, K., 2004, *SPIE*, **5165**, 457.

Martin, D., *et al.*, 2003, *SPIE*, **4841**, 805.

Martinez, P. & Klotz, A., 1998, *A Practical Guide to CCD Astronomy*, Cambridge University Press.

Massey, P., 1990, *NOAO Newsletter*, March, p. 16.

Massey, P. & Davis, L. E., 1992, *A User's Guide to Stellar CCD Photometry with IRAF*, NOAO publication.

Massey, P. & Jacoby, G., 1992, in *Astronomical CCD Observing and Reduction Techniques*, ASP Conference Series Vol. 23, ed. S. Howell, p. 240.

Massey, P., Strobel, K., Barnes, J., & Anderson, E., 1988, *Astrophys. J.*, **328**, 315.

Massey, P., Valdes, F., & Barnes, J., 1992, *A User's Guide to Reducing Slit Spectra with IRAF*, NOAO IRAF users' guides, NOAO Publications.

McGrath, R. D., 1981, *IEEE Transactions on Nuclear Science*, NS-28, No. 6.

McGraw, J. T., Angel, R., & Sargent, T., 1980, in *Applications of Digital Image Processing to Astronomy*, ed. D. A. Eliot, *Proc. SPIE*, pp. 20–28.

McLean, I. S., 1997a, *Electronic Imaging in Astronomy*, Wiley, Chapter 5.

McLean, I. S., 1997b, *Electronic Imaging in Astronomy*, Wiley, Chapter 7, p. 167.

McLean, I. S., 1997c, *Electronic Imaging in Astronomy*, Wiley, Chapter 12.

Meidinger, N., *et al.*, 2003, *SPIE*, **4851**, 243.

Meidinger, N., *et al.*, 2004a, *SPIE*, **5165**, 26.

Meidinger, N., *et al.*, 2004b, *MmSAI*, **75**, 551.

Meidinger, N., *et al.*, 2004c, *SPIE*, **5501**, 66.

Merline, W. & Howell, S. B., 1995, *Exp. Astron.*, **6**, 163.

Miller, J. S., Robinson, L. B., & Goodrich, R. W., 1988, in *Instrumentation for Ground-Based Optical Astronomy*, ed. L. B. Robinson, Springer-Verlag, p. 157.

Miyazaki, S., Sekiguchi, M., Imi, K., Okada, N., Nakata, F., & Komiyama, Y., in *Optical Astronomical Instrumentation*, ed. S. D'Odorico, *Proc. SPIE*, 1998, **3355**, 363.

Monet, D. G., 1992, in *Astronomical CCD Observing and Reduction Techniques*, ASP Conference Series Vol. 23, ed. S. Howell, p. 221.

Monet, D. G. & Dahn, C. C., 1983, *Astron. J.*, **88**, 1489.

Monet, D. G., Dahn, C. C., Vrba, F. J., *et al.*, 1991, *Astron. J.*, **103**, 638.

Mortara, L. & Fowler, A., 1981, in *Solid State Imagers for Astronomy*, *Proc. SPIE*, **290**, 28.

Nather, R. E. & Mukadam, A., 2004, *Ap. J.*, **605**, 846.

Neely, W. A. & Janesick, J., 1993, *Publ. Astron. Soc. Pac.*, **105**, 1330.

Newberry, M., 1991, *Publ. Astron. Soc. Pac.*, **103**, 122.

Oke, J. B., 1988, in *Instrumentation for Ground-Based Optical Astronomy*, ed. L. B. Robinson, Springer-Verlag, p. 172.

Oke, J. B. & Gunn, J., 1983, *Astrophys. J.*, **266**, 713.

Opal., C., 1988, in *Second Workshop on Improvements to Photometry*, ed. W. Borucki, NASA Conf. Publ. 10015, p. 179.

Peacock, A., *et al.*, 1996, *Nature*, **381**, 135.

Pecker, J.-C., 1970, *Space Observations*, Reidel.

Penny, A. & Dickens, R., 1986, *Mon. Nat. R. Astron. Soc.*, **220**, 845.

Perryman, M. A. C., Peacock, A., Rando, N., van Dordrecht, A., Videler, P., & Foden, C. L., 1994, in *Frontiers of Space and Ground-Based Astronomy*, eds. W. Wamsteker *et al.*, Kluwer, p. 537.

Pogge, R. W., 1992, in *Astronomical CCD Observing and Reduction Techniques*, ASP Conference Series Vol. 23, ed. S. Howell, p. 195.

Pirzkal, N., Pasquali, A., & Walsh, J. R., 2002, in *The 2002 HST Calibration Workshop*, p. 74.

Plucinsky, P., *et al.*, 2004, *SPIE*, **5488**, 251.

Queloz, D., 1995, in *New Developments in Array Technology and Applications*, eds. A. G. D. Philip, K. A. Janes, & A. R. Upgren, Kluwer, p. 221.

Rector, T., *et al.*, 2005, *A. J.*, in press.

Rieke, G., 1994, *Detection of Light from the Submillimeter to the Ultraviolet*, Cambridge University Press.

Reynolds, A., *et al.*, 2003, *A. & A.*, **400**, 1209.

Robinson, L. B. (ed.), 1988a, *Instrumentation for Ground-Based Optical Astronomy*, Springer-Verlag.

Robinson, L. B. (ed.), 1988b, *Instrumentation for Ground-Based Optical Astronomy*, Springer-Verlag, pp. 1–304.

Robinson, L. B. (ed.), 1988c, *Instrumentation for Ground-Based Optical Astronomy*, Springer-Verlag, pp. 189–304.

Roesler, F. L., *et al.*, 1982, *Astrophys. J.*, **259**, 900.

Romanishin, W., 2004, *Introduction to Astronomical Photometry*, http://observatory.ou.edu.

Rutten, R., Dhillon, V., & Horne, K., 1992, *Gemini*, Issue 38, p. 22.

Sabby, C., Coppi, P., & Oemler, A., 1998, *Publ. Astron. Soc. Pac.*, **110**, 1067.

Schaeffer, A., Varian, R., Cover, J., Janesick, J., & Bredthauer, R., 1990, in *CCDs in Astronomy*, ASP Conference Series Vol. 8, ed. G. H. Jacoby, p. 76.

Schectman, P. & Hiltner, W., 1976, *Publ. Astron. Soc. Pac.*, **88**, 960.

Schempp, W. V., 1990, in *CCDs in Astronomy*, ASP Conference Series Vol. 8, ed. G. H. Jacoby, p. 111.

Schmidt, M. & Gunn, J., 1986, *Astrophys. J.*, **310**, 518.

Schmidt, M., Schneider, D., & Gunn, J., 1986, *Astrophys. J.*, **306**, 411.

Schwope, A., *et al.*, 2004, ASP Conf. Series Vol. 315, p. 92.

Sembay, S., *et al.*, 2004, *SPIE*, **5488**, 264.

Shi, H., & Wang, G., 2004, *SPIE*, **5493**, 547.

Shiki, S, *et al.*, 2004, *Publ. Astron. Soc. Jpn.*, **54**, L19.

Skidmore, W., *et al.*, 2004, ASP Conf. Series, Vol. 190, p. 162.

Smith, M., 1990a, in *CCDs in Astronomy II: New Methods and Applications of CCD Technology*, eds. A. G. D. Philip, D. Hayes, & S. Adelman, L. Davis Press, p. 31.

Smith, R. M., 1990b, in *CCDs in Astronomy*, ASP Conference Series Vol. 8, ed. G. H. Jacoby, p. 153.

Srour, J., Hartmann, R., & Kitazaki, K., 1986, *IEEE Transactions on Nuclear Science*, NS-33, No. 6.

Sterken, C., 1995, in *New Developments in Array Technology and Applications*, eds. A. G. D. Philip, K. A. Janes, & A. R. Upgren, Kluwer, p. 131.

Strueder, L., *et al.*, 2002, *SPIE*, **4497**, 61.

Stetson, P., 1998, *Publ. Astron. Soc. Pac.*, **110**, 1448.

Stetson, P. B., 1987, *Publ. Astron. Soc. Pac.*, **99**, 191.

Stetson, P. B., 1992, *Publ. Astron. Soc. Pac.*, **102**, 932.

Stetson, P. B., Davis, L. E., & Crabtree, D. R., 1990, in *CCDs in Astronomy*, ASP Conference Series Vol. 8, ed. G. H. Jacoby, p. 289.

Stone, R. C., 1989, *Astron. J.*, **97**, 1227.

Stover, R. J., Brown, W., Gilmore, D., & Wei, M., 1995, in *New Developments in Array Technology and Applications*, eds. A. G. D. Philip, K. A. Janes, & A. R. Upgren, Kluwer, p. 19.

Timothy, J. G., 1988, in *Instrumentation for Ground-Based Optical Astronomy*, ed. L. B. Robinson, p. 516.

Tobin, W., 1993, in *Stellar Photometry – Current Techniques and Future Developments*, eds. C. J. Butler & I. Elliott, IAU Colloquium 136, Cambridge University Press, p. 304.

Tomaney, A. & Crotts, A., 1996, *A. J.*, **112**, 2872.

Tonry, J. L., *et al.*, 2004, in *Scientific Detectors for Astronomy*, eds. P. Amico, J. Beletic, & J. Beletic, 2004, Kluwer Academic Publishers, p. 385.

Tonry, J. L., *et al.*, 2002a, *Exp. Astron.*, **14**, 17.

Tonry, J. L., *et al.*, 2002b, *SPIE*, **4836**, 206.

Tonry, J., *et al.*, 2005, *Publ. Astron. Soc. Pac.*, **117**, 281.

Tonry, J. L., Burke, B., & Schechter, P., 1997, *Publ. Astron. Soc. Pac.*, **109**, 1154.

Trauger, J. T., 1990, in *CCDs in Astronomy*, ASP Conference Series Vol. 8, ed. G. H. Jacoby, p. 217.

Tüg, H., White, N. M., & Lockwood, G. W., 1977, *Astron. Astrophys.*, **61**, 679.

Tyson, J., 1990, in *CCDs in Astronomy*, ASP Conference Series Vol. 8, ed. G. H. Jacoby, p. 1.

Tyson, J. A. & Seitzer, P., 1988, *Astrophys. J.*, **335**, 552.

Vogt, S. & Penrod, G. D., 1988, in *Instrumentation for Ground-Based Optical Astronomy*, ed. L. B. Robinson, Springer-Verlag, p. 68.

Wagner, R. M., 1992, in *Astronomical CCD Observing and Reduction Techniques*, ASP Conference Series Vol. 23, ed. S. Howell, p. 160.

Walker, A., 1990, in *CCDs in Astronomy*, ASP Conference Series Vol. 8, ed. G. H. Jacoby, p. 319.

Walker, G., 1987, *Astronomical Observations*, Cambridge University Press.

Whitmore, B. & Heyer, I., 1998, *WFPC2 Instrument Report 97-08*, Space Telescope Science Institute Instrument Science Reports.

Wilson, C., *et al.*, 2002, in Low temperature detectors, *AIPC*, **605**, 15.

Woodhouse, D. *et. al.*, 2004, in *Scientific Detectors for Astronomy*, eds. P. Amico, J. Beletic, & J. Beletic, 2004, Kluwer Academic Publishers, p. 183.

Wright, J. & Mackay, C., 1981, in *Solid State Imagers for Astronomy*, eds. J. Geary & D. Latham, SPIE.

Young, A. T., 1974, in *Methods of Experimental Physics*, Vol. 12, Part A, ed. N. Carleton, Academic Press, Chap 3.

Zaritsky, D., Shectman, S., & Bredthauer, G., 1996, *Publ. Astron. Soc. Pac.*, **108**, 104.

Zebrun, K., Soszynski, I., & Wozniak, P., 2001, *Ac. A.*, **51**, 303.

Zhou, X., *et al.*, 2004, *A. J.*, **127**, 3642.

Index